Codex Orféo

Michael Charles Tobias

Codex Orféo

A Novel

Michael Charles Tobias
Los Angeles, CA, USA

ISBN 978-3-319-30621-6   ISBN 978-3-319-30622-3   (eBook)
DOI 10.1007/978-3-319-30622-3

Library of Congress Control Number: 2016937303

© Michael Charles Tobias 2017

This work is subject to copyright. All rights are reserved by the Publisher, whether the whole or part of the material is concerned, specifically the rights of translation, reprinting, reuse of illustrations, recitation, broadcasting, reproduction on microfilms or in any other physical way, and transmission or information storage and retrieval, electronic adaptation, computer software, or by similar or dissimilar methodology now known or hereafter developed.

The use of general descriptive names, registered names, trademarks, service marks, etc. in this publication does not imply, even in the absence of a specific statement, that such names are exempt from the relevant protective laws and regulations and therefore free for general use.

The publisher, the authors and the editors are safe to assume that the advice and information in this book are believed to be true and accurate at the date of publication. Neither the publisher nor the authors or the editors give a warranty, express or implied, with respect to the material contained herein or for any errors or omissions that may have been made.

Front cover picture: Wisent (ojwenzel/fotolia.com). *Back cover picture*: Snowfall (andreiuc88/fotolia.com)

Printed on acid-free paper

This Springer imprint is published by Springer Nature
The registered company is Springer International Publishing AG Switzerland

Dedicated to My Grandparents and Great Grandparents and To All Of Our Cousins and Loved Ones Who Survive In Our Memories

Contents

1. Prologue: Ivan The Great Bell Tower, Moscow, April 11, 2011 1
2. May 8, 2011, 12:25 pm, inside DS&T, McLean, VA 5
3. Two Hours Earlier, Near the Ladskie Artificial Lake 9
4. Nine Lives 11
5. Rio Centro, June 22, 2012 15
6. Equations for the End of the World 17
7. Nenhum dano não falta 19
8. A-Minor Key 21
9. Sasha's World 23
10. Stalker 27
11. Techsupport 29
12. The Other Alfred Wallace 33
13. Probability Errors in the Rhombicuboctahedron 35
14. What Would Truman Have Done? 37
15. A Room with a View 39
16. In and Out of Shadows 41
17. Fungus 43
18. An Uncommon Urgency 45
19. Eyes on the Ground, Spring 2012 47
20. The Road to Utopia 49
21. The Day of the Chicken 51
22. Deep in the Forest 55
23. Simon Stylites 59

24	Yakub Kolas Central Scientific Library	61
25	The Stampede for a Catalytic Agent	63
26	Thermus aquaticus	65
27	Alice's Rabbit Hole	69
28	Tuber aestivum	73
29	Pavilion 3	77
30	Rio Maelstrom	79
31	Euclid's Ghost	81
32	The Moment That Would Change a Lifetime	83
33	Ugly World	87
34	Daze	89
35	Obfuscations	91
36	Early Morning, Avenida President Wilson	99
37	The Bartender's Apartment	101
38	From the Knee Down	103
39	The Ides, 2015	107
40	The Communiqué	109
41	A Classified Huddle	111
42	noitnettaruoydeeniemsti	115
43	Cro-Magnon	117
44	Machinations	119
45	The Irish Connection	123
46	The Decision	127
47	London	131
48	A Glazed Twist	135
49	Warszawa Centralna	137
50	Pryvakzaĺnaja plošča, 9:07 am. Minsk	139
51	Vulitsa Kirava 13	143
52	Minsk Passazhirsky	145
53	No Birds	147
54	Unique Circumstances	149
55	Disaster	155
56	Scrimmage	159
57	The Great-aunt	161

58	Lev's Dream	163
59	225063 Kamieniuki. Region: Brest	167
60	The Histories	169
61	The Watch	175
62	The HIT	177
63	Convergences	179
64	The Coming Storm	183
65	Visions of the Shtetel	187
66	Twelve Feet of Snow	191
67	The Diary	195
68	Life and Death in the Forest	207
69	Despair in the Afternoon	211
70	Wise Men	213
71	The Graves	215
72	52°35′22.86″N 23°54′12.54″E. Elv. 201 M.	219
73	Eyes on the Ground	221
74	Plants That Dream at Night	223
75	Mystiques Amid the Rhizosphere	225
76	The Genius of a Squirrel	229
77	The Wisent of Bialowieza	235
78	Nothingness	239
79	The Marriage of Figaro Factor	243
80	Daybreak	249
81	11 a.m., Tenth Day	253
82	The "Parthenon" of Bialowieza	255
83	A Single Gnarled Gutteral	259
84	Escape from Nature	263
85	Paris	265
86	Les Deux Magots	269
87	The Synagogue de la rue Pavée, 4th District	273
88	The Survivors	275
89	The Crossing Over	279
90	Coda	281

© M.C. Tobias

Chapter 1
Prologue: Ivan The Great Bell Tower, Moscow, April 11, 2011

The sturdy, early middle-aged American, his diplomatic visa strategically tucked in the old-style tactical Velcro, flat against his t-shirt and taught abdominal muscles beneath winter garb, climbed the endless steps up into the public armpit beneath the 70-ton Assumption Bell in the arch of the belfry.

Surrounding him were nearly two-dozen bells ringing all across the Kremlin and throughout downtown Moscow, emanating from the various cupolas of the Ivan The Great Bell Tower complex. Rays of sun shone through dark, fast-moving clouds.

The Eastern Orthodox tradition of the bells was startling, up close, as well as beautiful. John Vespers, ironically, appreciated the clamor for more practical reasons on this day. The din was precisely the point of this hastily arranged meeting.

A lanky kid in frighteningly good shape, mid-twenties, in tight-fitting jeans, a leather jacket and woolen scarf, tennis shoes-slash-hiking boots, directed his one word query to no one in particular: "You?"

The early middle-aged American nodded. He resembled more or less every other tourist to Moscow's most famous complex of churches during a season that was still, by all standards, freezing. He'd already had the location fully scoped out for line of sight and video surveillance.

"So?" asked the American.

The young man shook his head just slightly to indicate "*Nyet.*"

"How unfortunate. For you."

The words took the White Russian by surprise.

"You know about the crash?"

"Uh-huh."

"At Smolensk. The Polish president was killed, many top military advisers."

"Yep."

"So. You know about it."

"Everybody knows."

"There was a fireball."

"Right."

"Rain, sleet, fog, chaos. Military closed off the area for miles around."

The American said nothing. Waiting.

"Everything had been prepared. The one viable sample on board that aircraft, the only one I've ever managed to secure."

The American looked at his watch, then glanced away.

"In five hours, more or less," he said in a cynical declarative, "three Western Lowland gorillas in the Cameroon, a critically endangered species, will be injected with a serum. But the serum will fail, as it has been explained to me. Not my area of expertise, you understand. Anyway, the animal's genome was thought to be in the neighborhood of 3,000 megabases, millions of nucleotide base pairs – give or take. You probably understand all that. But the animals are going to die. And seventy million human beings will also die, this year, for any number of reasons. Cancer. Cardiovascular. Dehydration. Depression. And, of course, old age. Alzheimer's. So you. I am no Mother Teresa, but you *are* the egghead. And paid well, are you not?"

"*Da.*"

"You are at ground zero, no?"

The contact was utterly off his guard. Two days before, he'd been in the forest. In a storm that dumped over five feet of snow in 36 hours. But it was not only the weather that had, during the past two years, plunged dramatically into a cycle some were labeling D-O, a Dansgaard-Oeschger oscillating event. That was not what made him so nervous, even frightened.

Something else had happened out there. He couldn't mention it or describe it, even if he tried.

Moreover, the contact did not expect the onslaught, and certainly not from an American official of high standing, in so exposed a location, the ambient confusion of sounds notwithstanding.

"We're out of time here. Tick-tock, tick-tock."

"I'm telling you." Now the young man was fuming. "The crows, the insects, the breakdown is rapid, and so much ice."

The American studied the contact with a mixture of incredulity and bathos. "I had high hopes for you."

Visibly angered, the contact looked around nervously. He knew his every word might be interpolated visually by a drone, or a camera positioned anywhere within a mile of the Kremlin, no matter how careful his preparations for the meeting. But it was the monthly cash that now seemed in total jeopardy.

"Nobody can hear us," the American said. "We're standing precisely where we need to be."

The contact looked down and mumbled under the frost emitted from his breath. "The snow starts turning to rain in the forest. We're in a peculiar zone. The temperature variations are easily sixty, seventy degrees in twenty-four hours. Eight feet of snow on the ground at present. That makes sampling next to impossible."

The American drew in an ugly, compromised breath through half-choked nostrils, calming himself.

"You don't understand," said the young man.

"Really? What don't I understand?"

1 Prologue: Ivan The Great Bell Tower, Moscow, April 11, 2011

"Something's changed. I can't explain it."

"Really."

The contact could not work out the conflict within. He started to say something, then thought better of it.

"Visitors? To your field station. There are visitors from time to time?" He more than knew that there were.

"Very rare."

"Really. You have siblings? Other than your late brother, of course. A girlfriend who visits?"

"You're threatening me."

The American turned again to admire the view. "Don't go getting paranoid on me, now."

The student eyed his boot as he fidgeted around to rub out an old cigarette butt on the stone path. He almost seemed to be counting out beats in his head. Actually, it was the money he knew his mother would soon be without.

"How hard can it be to extract another sample from beneath the snow?"

"I don't know."

"Say again?" the American said, dismissively.

The bells were now clanging in unison, hundreds of tons of iron, copper, and steel from the Church of St. John Climacus and the Cathedrals of the Archangel, Assumption, and Annunciation, sending their explosive harmonies out into the afternoon chill.

A group of tourists passed by the two men who turned, leaning against a railing as if to admire the glories of ancient Moscow in the spring. Two nuns carefully avoided them, almost as if they were giant insects in the way of an important pilgrimage.

The American's impatience peaked. "Listen to me very carefully. Two weeks."

He handed him a slip of paper with two telephone numbers written on it.

The young man, sullen and resolved, nodded with a distinct ire, and read the paper: "Skifosovsky Institute of Emergency Care, 6806722; and #103 for general medical emergencies."

"I hope my handwriting is legible."

The young man took off and disappeared into the crowds.

Chapter 2
May 8, 2011, 12:25 pm, inside DS&T, McLean, VA

Jake Cosgrove, a former remote sensing geek from NASA, had joined the Agency while still in his late 20s. Now, just shy of his 30th birthday, he was the go-to guy when it came to anomalies.

Usually, that would entail the manipulation of two dozen different high-security image flows from the 528 or so U.S. satellites, as well as a barrage of other data – shared, or not – from the more than 737 other satellites circling the planet. New algorithms enabling Cosgrove's team to capture real-time data proved, at the very least, helpful in obtaining what was in his and his peers' deep computer bunker vernacular, either a "single," "double," or "triple whopper."

The younger Claire Upton, a graduate of MIT who worked down the hall and specialized in rural Eastern European GIS, an odd confederacy of visual fogs, blurs, and not infrequent eruptions, buzzed her senior colleague.

"I need you."

Jake swiveled in his chair amid his ridiculous array of computer screens, past his little 85 mm glass snow ball with the customary engraving on the brass base, *Veritatem Cognoscere*, We Know the Truth, and picked up a his cell phone, which was now also buzzing.

"Hey, sexy."

"Shut up. We'll both be fired."

"Never. They rely on us too much," he said, grinning boyishly. Jake sported a tiny platinum earring in his left ear, rarely discernable under his long mop of dark hair. He wore jeans to work. In his faraway cubicle, nobody bothered him. Claire, a freckled towhead, looked positively underage to be working at the Agency. The two had been dating for over a year.

"So what's up?" he asked.

Rather than make the forty-yard trip down the hall, he awaited her internal memo. It came over the sub-network and instantly downloaded itself onto his mainframes – all four of them.

"Okay," he mused, scanning first the EO-1 10-meter multispectral image, then an Enhanced Thematic Mapper Plus 15-meter image from Landsat 7, as well as two parallel blow-ups of something odd from TerraLook's Spaceborne Thermal Emissions and its ASTER Reflection Radiometric data.

"Looks like smoke, mirrors, and some nasty action down there."

"Lightning strikes in the snow, a passing shower, and yes, that is gunfire, three men, three dogs off their leashes in clumsy but zealous pursuit, and the area has been slammed by more than five feet of snow just this month," Claire stated.

"Are those guys military?" Jake questioned. "Resolve. Zoom in. Okay, just a sec…uh, there's cows."

"Actually, those would be wisent, like our bison. I count nine of them," Claire corrected him.

"It's happening right now."

"Well, maybe a three-second delay."

"Where are we…wait, I got it." He was fast scanning all the incoming data that came with the image uplinks. "Right. Belarus. The last dictatorship in Europe. Looks kinda pretty, actually."

"It's a National Park, bozo. 'Course it's pretty."

"Looks freezing. Uh-huh. There he is."

"Minus nine at the moment."

"Oops! Down we go. Shit, but not entirely."

They both watched on their screens as a shadow figure crawled tortuously through the deep snow, in and out of dark space – the effect of the resolution compromises and/or shadows – dragging something, with the three humans in sluggish pursuit, dogs desperately clawing through the heavy snow. The men had discharged at least three bullets from long rifles into what appeared to be the victim's right and left shoulders, and possibly clear through his left tibia, based upon the blood-splatter effect, which pixilated inordinately on their screens.

"Oh, dear," she whispered.

"Go back," Jake said.

Claire froze the image, spun it around, then replayed from the opposite direction.

This time they almost, but not quite, could make out the real man, definitely the agony, and the very speed at which the bullets sent him into the air, in between the long, looming darkness laid down by the enormous trees at the end of the day.

"AN-94 rifles," Jake offered.

"That's what they use there."

"Is that a nest? What is that, a bear?" Jake was catching something high in one of the trees as he scanned wider.

"Possibly an eagle's nest. Now look closely."

She zoomed in on the truck about 2,500 feet away, parked beside a lake. "I've been following it for two hours. Military, already got the license."

"Okay, so? What's up? I can't remember. Do we even have diplomatic relations after that Senate Human Rights Bill, remember?"

"Of course we do, strained but…You know that great trattoria just off New Hampshire Avenue, serves that killer chopped salad and gnocci mac? Your parents took us there."

"Right. Love it."

2 May 8, 2011, 12:25 pm, inside DS&T, McLean, VA

"It's just a few blocks away from the Belarus embassy. You're eating."
"Hummus and felafel wrap."
"Don't you ever get tired of it?"
"Nope. I've saved half."
"Get over here."

Jake walked across to Claire's office, certainly the most pleasant thing he'd done all morning.

Immediately, she took a bite of the wrap, then pointed to three of her computer screens that highlighted, in different levels of resolution and color correction, the man who'd toppled onto the snow, and the pack he was wearing.

"Right, so? What's up? He's basically in and out of the shadows."
"Hundred-fifty-foot high Norway spruce, I believe. Oldest forest left in Europe. It's 20 minutes before their sunset. Long shadows in a storm. Hard to read. *Verst* number 3, that I know for certain."
"What does that mean?"
"I believe it's one of many highly restricted zones. They're measured in versts."
"The whole country's restricted."
"Yeah. But this place is doubly so, like off-limits to nearly everybody."
"Power plant nearby? A prison?"
"No, no. I told you. It's a National Park. But this guy obviously shouldn't have tried his luck."
"National Park, but off limits? I don't follow," he admitted.
"One of those strange parks. Nobody ever goes there, not on the eastern side, hardly ever. Just scientists, and – from the looks of it – the occasional poacher, or whatever. There is a high military presence smack down the middle. The park is both eastern and western sectors, two countries, Poland and Belarus."

They watched as the three men arrived at the site of their fallen prey, along with the converging dogs that they held back from the poor bastard, and proceeded to force him to his feet, rifling through the backpack as the man collapsed again. They let him fall, and the blood poured from three points in his trousers.

"Jesus…" Claire groaned.

The soldiers, in high boots and long coats, stood over the figure sprawled on the ground, as the sunlight bore down on the freaky scene admitting openings through the passing storm. Without those openings, most GIS was ineffective.

"That would be hail leaving such an imprint on the snow, no?" Jake aired aloud.
"Probably."
"I guess they're discussing the guy's fate."
"I'd say it does not look good."
"No," Jake concluded.
"So what brings you to this unfortunate fellow's plight?" Jake asked.
"A shitload of noise, between a few embassies, the Belarus KGB, but also – you'll love this – our State Department."
"Henry Lew's office?" Jake wondered out loud. Both he and Claire knew him socially, loved his wife's Chinese cooking. "I like Henry."
"I don't think so."

© M.C. Tobias

Chapter 3
Two Hours Earlier, Near the Ladskie Artificial Lake

Aleksey Artemios knew the sound with a level of fear that traveled like blood through his veins. Strikes of lightning and the subsequent rapid thunder. Twelve years before, he'd been among the lucky ones to survive the Nyamiha disaster in Minsk when a freak storm terrified hundreds of people in one of the city's crowded metro stations. There was a stampede for cover, everyone slipping on the pavement, and more than fifty people were crushed to death in a tunnel.

Now, as he made his way back towards his encampment, he simply chose to wait it out under cover of a huge forest grove. The trees had formed a vast network of exposed roots, and some of the huge icicles had engendered a cave-like opening in which he took shelter. It offered a surprisingly high amount of heat.

His residual fear passed with the storm, and he now prided himself on his catch: several pounds of a certain fungus that only grew, as far as he knew, in this one small corner of the forest. His grandfather had discovered the area fifty years before, at great peril to himself and family, and since that time, the so-called "winter mushroom" had become all but inaccessible, if not nearly extinct in the wild.

Aleksey knew there were terrible risks, but he also now benefitted from something that would have been unimaginable to Aleksey's grandfather: it turned out that this particular fungus, which thrived under deep snow, contained certain chemicals, and a pharmaceutical company was happy to pay a fortune for whatever Aleksey could unearth.

It was a bonanza. In his wildest dreams he could never make such money from any normal activity.

As he resumed his trek back through the deep snow, he heard the one sound he had prayed he'd never have to worry about: barking dogs!

Perhaps they belonged to some other hunter, illegally doing his thing. That he could easily survive; he had in the past.

But then…His fate was sealed without question. They were Transylvanian hounds, the cries of their masters goading them on.

It was all over for him. He sat down beneath the nearest tree.

Chapter 4
Nine Lives

An explosion shattered his already convoluted dreams... It was not the first of such detonations. A lifetime of bad news had conditioned him to the iconic fireball that could mean so many things.

Still, it was powerful enough to wake him from a snooze.

The after-image was not one he particularly cottoned to at 39,000 feet. Professor David Lev had fallen asleep, briefly, while reading through his notes to his assistant, Malcolm Howler – Mal, for short – a brainy little nerd who was getting his doctorate under Lev. They were in Business Class en route to Rio. Mal had never flown Business Class, nor had Lev; he typically flew First Class, unapologetically. We're the ones paying the CO_2 tax, he'd say.

In fact, it was an inoperable back problem he'd had ever since jumping off a 120-foot cliff to escape being eaten by a mother grizzly bear in Alaska, many decades before, and his hypertension – two beta blockers a day kept it relatively in check, but also a recent situation, erratic and severe panic attacks that he fought off with a combination of drugs and walking back and forth – all of which, including prostate issues, made long-distance travel very difficult and painful for him.

Neither his doctor nor psychiatrist could explain Lev's panic attacks other than to suggest that they can hit people at any time in their lives. In the professor's case, a lifetime of cumulative insults to his body from his often outrageous field research on every continent, two plane crashes, including a very ugly one in Yemen, his being seconds shy of perishing in an avalanche in Chile, severe heat prostration in both the jungle of Suriname and spiny desert of Madagascar, etc., – not to mention a childhood with heart-shattering components.

Lev sat uncomfortably reading off his computer, racing through his bullet points for his upcoming speech. "Here's how I propose opening. Consider the following scenario: You and your fellow hikers are hanging around Mt. Everest base camp. A spectacular storm is coming in from Tibet. It's the greatest view on earth, although you've been warned about collapsing seracs up ahead in the notorious Khumbu icefall."

"You're wearing earplugs listening to Beethoven's Ninth on your iPad, devouring oyster mushrooms in a sautéed orange and tofu sauce. Your head hurts. That's normal. You're peeing a lot, also normal. Chatting up the rhododendrons that were in early bloom around Namche Bazaar, several easy days out of Kathmandu. The Fourth Movement. It doesn't matter who's conducting – Otto Klemperer, Wilhelm Furtwängler – somewhere between 23 and 29 minutes, depending on the pace of the baton. That's how long it should be."

"Seems to me you've already risked boring them or losing them altogether," Howler unhesitatingly chimed in. "Just being frank."

"Hold on. The kicker's coming. Like some nasty little Brothers Grimm fairy tale far to the South in the Indian Ocean, indeed, in all the oceans, it starts." And he wriggled his fingers and went "ooooh!" like it was some scary shit coming.

"Not your microns?"

"What's wrong with my microns?"

"We went through this. You agreed it was too esoteric."

"I shortened it. A trigger effect. Bingo! The most numerous life form on the planet, microbes of the genus *Prochlorococcus*, about 3-times-ten-to-the 27th – that's measured in octillions – goes dead on us. A tiny microbe."

"I'm telling you. You've already stopped them in their tracks with two utterly unknown words and one totally esoteric measurement – a very feckless accomplishment for all of one sentence. How long was your speech?"

"You asshole."

"Professor, look at me. I'm your audience. I'm trying to help. This is a plenary session for the UN."

"I know precisely what it is. I was at Kyoto, remember? I wrote many of the clauses integrated into the final report, remember?"

"I don't exactly remember. I was six years old at the time reading the Hardy Boys. But I do know Kyoto was not binding. And I also think you're turning this, what could well be your last chance, let's be real, into some sci-fi grade B movie. Or the Apocalypse. The world's not going to end that fast. It'll go out like, I forget how T.S. Eliot put it…a whimper, not a bang…something like that. Also, people hate being preached to, or told what to do."

"My specialties. And 'April is the cruelest month,' Eliot said. Which is totally the point. You ever hear of the Hundred Years War, Four Black Plagues, and now, the worst drought in a thousand years in your beloved Yosemite? I know what I'm doing."

"Prof, you wanted my opinion, you got it. Everybody believes they have nine lives. It's like what you once described in class, about that dude Jacob Bernoulli, the Swiss mathematician who died in the early 1700s."

"1705."

"Of course you'd remember that precisely. My point. Anyway, he left, as you described it, a most telling motto behind for his gravestone at some church."

"The Basel Munster."

"Can I finish? 'Changed and yet the same, I rise again.'"

"Good memory. *Eadem mutata resurgo*. But Mal, you silly child, point two microns give or take, producing about fifty percent of all the oxygen on earth is hardly esoteric."

"Yeah, it is."

"You kids are very annoying, you know that? I confess to utter dismay."

"Uh-huh," Howler grinned self-mockingly. Lev called all his students "kids," – or worse.

"You're up there at Everest base, oblivious to these goings-on, at about 18,000 feet, which means you've got half the normal oxygen. Half. Get the picture? I threw that in for a twist, and I think it's quite clever. Of course, in your case, half would mean more like eighty percent."

"You're really good, more like an ecological sadist."

"Heh, sorry, but those of you at Everest base camp are going to enjoy, at best, one minute of "Ode to Joy." *Auf wiedersehen, Ludwig*! I think it drives the point home quite clearly."

"You don't even have an Everest or microbe image on your power point," Mal reminded the old professor, restlessly shifting his own materials, spilling a glass of red wine everywhere.

"Good work, kid," Lev said, wiping the wine off his computer.

David Lev paused, putting down his pen as the plane lurched, and looked to his grunt. "These are presidents, tribal chiefs, prime ministers. They need to know what's coming."

"You really think they haven't connected the dots, all those grim biospheric coefficients in one gigantic, contaminated beaker?"

"No. I don't. If they did, they wouldn't keep going to these global summits."

"They'll do anything for mileage."

Mal loved screwing around with his cantankerous guru whom he had admired for half his life. Lev was the reason Mal applied to UCLA.

He'd first read the famed professor's early papers on quantum biological systematics while still in high school, when his dream in life was to be involved in helping to create National Parks around the world, while his peers were all focused on Silicon Valley or Wall Street.

Lev grinned. He knew the kid was right. But he also knew that human understanding was terribly flawed, in the face of whole civilizations, comprised largely of young people Mal's age willing to elect leaders who prided themselves on never having read a book, as Joseph Roth once described President Hindenberg. Of a generation of nerds better equipped to take selfies than read John Milton.

Lev and his wife of nearly half a century, Sasha, had chosen not to have children like Malcolm, not that he could fault the 26-year-old for anything other than being a sterling prodigy for whom Lev felt certain paternal instincts; but Lev had been teaching other students at various universities for decades and most of his colleagues had grandchildren, even great-grandchildren, and he'd heard many of those little brains of entitlement dimly glossing over the highlights of animals going extinct, of our singular species putting its affairs – not in order, but disarray.

"I need some shut-eye," Lev said, searching for a few pills from his black Prada zipper vest, grabbing hold of a wine glass and taking a swig of the remaining Cabernet. Then he wrapped his own goose down pillow around both sides of his head, and put on his eye mask.

Chapter 5
Rio Centro, June 22, 2012

E6›n1^%/∞ flashed on the screen… *take your goddamned pills… hurry.* Lev panicked…

Low-flying helicopter flotillas raked the sky, ominous drones streaking here and there, while morning showers gusted through the arriving human mobs beneath the coming storm clouds.

***

An hour before…

David X. Lev stepped into an open seat in an electric golf cart at Portão 1A off Av. Abraão Jabour.

Alongside his trusted protégé, Mal, who carried the heavy briefcase, they braved a nearby swarm of some 15,000 paramilitary security forces, all bearing shields and heavy machine gun fire power, acrid spray-throwing devices, like Roman legions or Republican guards, generally directed toward the clusters of protestors all kept back by a barrage of barricades.

The cart carried four people plus the driver, racing off at a brisk five kilometers per hour, through the maze of the Convention Center hosting the United Nations Conference on Sustainable Development, meant to be a fruitful, sleeves-rolled-up pragmatic paen to the earth, in the region of Barra da Tijuca, 40 minutes from Lev's hotel just off the *balneario*, the famed four-kilometer stretch of Copacabana and Ipanema beaches. Mal's cot was in the living room.

The particular bairro area in which Lev and his grad student were staying for four nights used to be known as "Sacopenapã" in the local Indian Tupi language. Lev had read on the plane that it roughly translated to "the way of the socós," or "*socós boi*," today known as the Rufescent Tiger Heron (*Tigrisoma lineatum*) of the Ardeidae family, details that all meant something to the famed Professor Lev.

There was one continuous traffic jam out to Rio Centro. Inside their taxi, Lev tried to control an unstoppably runny nose and that telltale itch in the throat.

"Did you bring any Robitussin?" Mal asked him.

"No." He opened his little iPhone and dialed hastily, whispering to the message machine, "Sweetheart, I know you're asleep but I just wanted to let you know I love you…" He reflected silently, "Everything's good here. But I'm never doing this again. Glad I brought the kid along, though."

He pocketed the apparatus, already two generations out of date, and gazed into the scrambled rush of his crunched data and typos, all accumulating into so many acronyms and a tacit shorthand for doomsday.

With the typhoon of conferees, the broadband was breaking down with exasperating predictability. The exhausted 84-year-old Lev had spent half the night rewriting his twenty-minute speech by taking various salient cues from utterly depressing presentations he had endured thus far, during two days of non-stop 8:30 a.m. to 6 p.m. deliveries and proclamations, private meetings, and "high-level sessions."

Everyone concluded nervously optimistic briefs and diplomatically declared all the right things, with every reassurance that humanity had what it took to get it right.

I'm going insane, Lev thought, buzzing, hyperkinetic.

No one seemed to share Lev's own Petri dish-like Tonic-clonic, a grand mal dizziness of metabolic gymnastics on the dark side, as equated by kilowatts, kilocalories, billions of hungry human mouths, and biochemistries of sleepless nights. But then, the legendary professor had been far beyond any level of comfort for decades.

Instead, what he and Howler had deciphered in the slew of headlines was the parading of one insulting simplification of global demise after another, forcing Lev to re-phrase his own re-balancing act depicting the severest of puncture wounds, coral blanching, and tree felling in anticipation of one hell of an audience with real-time telecasting throughout the nearly mile-square Rio Centro and its contiguous pavilions.

"You'll be fine," Mal declared fearlessly.

"Thanks," Lev said, humbly appraising the stakes before him.

Lev knew this would likely be his personal farewell. Mal was right; the final keynote of an arduous, and no less cynical, trajectory, teeming with death knells, certainly the speech with the greatest bad odds looming over its heart-felt content, not to mention the venue itself: a globe-studded scrabble board of consequential question marks.

"You need to be Socrates to counter their comfortable piña coladas," Mal waxed in his stride the previous night, supping on crispy quinoa tapas in their rusticated little boutique hotel's neighborhood.

"You're missing some great shrimp," Lev teased his younger, vegetarian companion.

"After your upcoming speech? How can you even *think* like that? You're about to tell the world the seas are dying."

"In the next life I'll try to become a vegetarian. Vegan, probably not. I love good French cheeses too much."

Chapter 6
Equations for the End of the World

They arrived at Pavilion 5. The golf cart went no further, letting off its passengers and zipping back to Pavilion 1 to start its rounds all over again.

Stepping off and merging into the chaos, Lev, with Mal by his side, moved past one of countless journalist hubs – huge arrays of dedicated computers, television crews, reporters doing live coverage, interviews occurring every 10 steps.

Their ID badges – vouchsafed many months before by the United Nations Accreditation Protocol Unit – were triple-scanned. That was the procedure for the thousands of delegates. At length, they arrived at the podium where Lev was to give his speech.

Observing that there were four arriving prime ministers and presidents at his table and Mal seated next to him, Lev was feeling the electricity in the air. Possibly as many as three thousand people in the hall, some in headfeathers and little else.

"... *don't say shit or fuck*"... Lev reminded himself – he had to be careful to do that – taking his seat, as well as opening the water bottle in front of him and downing a quick cocktail of pills to stave off his panic attacks, all with surreptitious efficiency, notwithstanding video cameras everywhere, in every shape and size.

"Dr. David Lev," a voice declared.

It was the professor's turn. No one in his profession knew of his middle name, always the initial, "X." as he never used it. Only his wife, of course, was aware of its peculiar presence, following upon – not the marked "spot," rather – the Star of David, in a most precarious family history. That and, according to David who read into his parents' frighteningly anticipatory choice: Exiled.

E6 ›n1^%/∞ lingering without the slightest meaning on the giant screen…

Only a few in the audience would know of him for his modern version of the celebrated Ehrlich-Holdren formula, I=PAT (impact on the environment is the equivalent of population times affluence times technology). But he honed in to focus expressly on the mathematical attrition of populations and species in light of our consumption within a so-called "*tragedy of the commons*" scenario. Namely, E6 ›n1^%/∞; where E6, the now current Anthropocene of sixth-generation biological extinctions could not be disassociated from the branching stochastic mathematical function signifying that the E6 event must necessarily result from n equaling 1 multiplied by the Zeno Paradox of *H. sapiens* fertility rates. That would only resolve at the point of infinity, and not before, regardless of the ratio of fertility to population size.

In common parlance, the world was in serious trouble, and basic arithmetic proved it.

Mal was concerned. He knew his guru had already lost the audience.

Lev clumsily clicked his Power Point and the enormous screen lit up with slide two, the familiar PETM: Paleocene-Eocene Thermal Maximum. A pale blue background, and a somewhat silly montage of dinosaurs that Malcolm had thrown together for the professor, did little to invite much interest in the words themselves, or the 200,000-year saga of a 5°C (9°F) globally transfixing temperature increase some 54 million years ago.

Amid the rash of temperature and salinity data, Malcolm was desperate for the professor's Mt. Everest story, as he scanned the packed auditorium to gauge reactions; anything to get the mob galvanized.

Lev sent a discreet query in passing to Mal, whose scarcely concealed grimace seemed to convey the negative Pap smear of public opinion. Lev sped it up.

Chapter 7
Nenhum dano não falta

Suddenly there was a commotion from the back of the huge assembly, with people scrambling to all sides.

Without warning, a very real-looking hatchet came flying at high speed above the heads of over a thousand people, smack through the middle of myriad journalists – expert aim, the best – passing five tripod-bearing television cameras and people in chairs, only to sink with perfection into the dead center of a map of the Amazon that had been trotted out, glued on something like a 5 ft. × 5 ft. posterboard, to the side of the round table, the compliments of Brazil's Indigenous Peoples Administration, showing where several hundred un-contacted tribes were still being protected.

"Fuck me!" Mal mumbled. "A tomahawk!" He looked anxiously at his professor, who wasn't quite sure what was really happening.

Bodyguards lunged at their marks to protect heads of state at all costs, knocking the notes of various speakers off the table. A water glass shattered on the floor. Two presidents were now kneeling while their security details raced to apprehend the assailant. Sniffing dogs moved with whipped-up frenzy through the assemblage.

Within minutes the ruckus was over. A single Indian, the representative of his tribe, had been wrestled to his knees and handcuffed. Dogs nipped at his exposed flesh.

"Now what?" Lev asked the moderator, as both men stood to their feet.

"Ladies and Gentlemen! Ladies and Gentlemen," the Moderator shouted above the noise. "Please understand, everything is fine. The man from I believe the Upper Xingu area of Amazonia was allowed to carry objects of his culture's sustainable tradition through Rio Centro's magnetometers."

Upon rapid scrutiny by converging police, however, the hatchet – which had looked to be of wood – turned out to be a tomahawk fashioned from more or less harmless plastic.

The weapon was removed from the map up front by guards. It was handed to the moderator, who held it up high for all to see.

"Neither spear nor dart gun nor anything made of metal. It is a play tomahawk. Plastic. A toy," he laughed nervously. "A toy, I repeat. No harm, no foul. *Nenhum dano não falta*. The gentleman has made his point, and without injury, blessed be God. And we are all the more enlightened for his welcome commentary."

Presidents and prime ministers re-took their seats.

Now Lev – who was wondering *why the fuck would a tribal chief take issue with my equations* – was operationally back in business at the podium, poised to wrap it up. He had three minutes.

Chapter 8
A-Minor Key

"On that appropriate note, ladies and gentlemen, let me just conclude by saying that it is not just Amazonia where destruction is spiking to the tune of 7,000 square kilometers annually, a size greater than Belgium. Our species is consuming some 70% of the Earth's biomass."

Lev's wife, Sasha, never took her husband's sweeping desolations too seriously. She'd been through too much in her life to be ultimately concerned. She knew very well how to survive, in her own time, in her own way. And, frankly, so did Lev, despite his myriad jeremiads.

Indeed, "In summary," declared Lev, "all is not lost."

He referenced the few bright lights in his world, places like the nearby Guiana Shield, particularly the bulk of Suriname; the democratic nation of Bhutan; and the Rumanian/Ukrainian Danube Delta, the second largest and best preserved of riparian forests and habitat in Europe.

It was then than he began to feel the telltale signs: his feet were freezing, his head spinning – all the symptoms of a serious panic attack.

As Mal helped Lev to his seat, amid muted applause from the audience, Lev, sipping water, downing a few more pills, flashed into his own zone…

…*towards some horizonless azure light. At the near-edge of a world given to springtime. A little hand grasped at a buttercup. Far off, massive bison skirted the fringes of the forest… Paleolithic behemoths stared directly at him, and at the child by his side; mist from the animals' huge nostrils, heavy-hoofed, the masters of the forest who defined the childhood's greatest wonderments and humility. The percussion of their footsteps unheard, but the earth seemed to rumble. Even then, a kind of nostalgia… those foggy mornings… smoke rising from chimneys in one farmhouse after another… but also, the tension in the air… the inevitability of fearsome change…*

"You all right, Professor?" Mal asked in a whisper.

"Yeah, I'll be all right."

"Listen, something's happened back in L.A. I haven't picked up. But there's a jumble of weird text messages on all our devices."

© M.C. Tobias

Chapter 9
Sasha's World

West Los Angeles, 6 a.m. Sasha Lev listened to her husband's message. She could tell by his voice that he was not feeling well, and she worried with good reason: he had been rushed to the ER four times in the past 24 months. The symptoms were diverse, but all seemed to hasten the notion of a looming heart attack or stroke.

His panic attacks clouded the overall assessment of his medical specs, and no one could quite account for them, except Lev himself, who never uttered a word of it, although the internal landscape of fire alarms and pitfall traps was an unambiguous picture of his historic, unexamined life, just as X. never surfaced in any conversation, publication, or even the most private of discussions between Lev and his spouse, or among the few very close friends with whom the Levs kept occasional company.

Sasha obeyed her routines. Yoga most mornings at around 5:30. Early riser. Coffee and fruit plate, a piece of whole grain toast and a generous spreading of crunchy peanut butter while she read the New York Times, the L.A. Times, and the Economist.

Feet propped up on the inherited Wiltern Knit wool couch whose pillows she had years before hand crocheted in a subtly beaded design after photos David had once taken of Tibetan black-necked cranes during their migration through Central Bhutan.

She suffered from two hairline fractures in her left foot, choosing not to have the operation, but rather keep little pads in her desperately narrow-sized tennis shoes and thereby manage her daily hikes, and thrice-daily swims in their duck-shaped little swimming pool out front. It was hidden by a huge wooden wall, covered in trumpet vines that occasionally succumbed to whitefly, but her gardening team from somewhere, once, in Mexico, stayed on top of it. The budget for all this was guaranteed by her husband's tenure at UCLA, although inflation and a completely idiotic approach to the market (on her husband's part) had dented their principal quite substantially.

Still, there was always the house that could be sold, should they really screw up in terms of responsible planning for old age, which they were well into. She always mocked the concept, knowing – particularly during the past few years – that both she and her husband had already far outlived the actuaries.

Indeed, given their respective Jewish itineraries throughout an unfathomable twentieth century – for Jews, certainly – she thought it nothing less than miraculous that they were both still alive, still married, and – everybody's guess – still happy.

It had all gone by so quickly... a wondrous cliché she related to, considering the phantasm of World War II, which she had endured in a London subway, and the fears that came with the Cuban missile crisis, something one never forgot – how close, how uncanny the potential for repeat, and how vague the reality that there were still nearly 16,000 nuclear warheads sitting neat and tidy and all ready to be launched in at least 9 nations, some emerging with fanatical alacrity and hatred of the Other.

But David's phone message underscored her happiness this day, her having long absorbed life's intolerable crises to focus, instead, on its small pleasant gifts. She felt fortunate, even blessed.

At least until the phone call at 7:03 in the morning.

"Yes?"

The voice, a youthful one: "Mrs. Lev? You don't know me, although we've met once. On campus. I'm Josh Balfour. I work in your husband's lab. Sorry if I woke you."

"I'm up, Josh. What's wrong?"

"That's just it. A bunch of strange spam. Ordinarily, I would not have thought twice. But in this case, well, I had to open them because I'm not quite sure what to do."

"Go on." Her heart was starting to pound.

"My close friend, Malcolm, Mal, he's with your husband in Rio."

"Correct."

"Well, I just opened like six messages that, in combination, are just too weird." Pause. "Are you there?"

"Yes?"

"Encrypted stuff. Not my forté."

Silence.

"Mrs. Lev?"

"Yes. You've got my full attention."

"I'm sorry, Ma'am. It looks like – definitely a hacker."

"So what is the big deal? We all get spam. Yesterday someone claiming to be the Vice-President of a Nigerian bank invited me to go into business and make a bundle. Another offered me advice on erections, hardly relevant, at my age. But what of it?"

"This stuff is just weird. These spams know too much."

"Go on?"

"'KS.' Then '*iber azhentsy.*' Also the word '*Prometej.*'"

"I have no idea what any of this means. Do you?"

"No. They were sent to Malcolm Howler. His full name. We went to Harvard West Lake together. Close friends since junior high."

"So what is the point? Why are you calling me?"

"There was an address in Rio."

"The address of the conference or their hotel?"

Sasha could hear the submarine-like "ping" of another e-mail on Josh's end.

"No. Ooops – there's another one. It's some downtown Rio district. Nowhere near their hotel or Rio Centro. But here's the freaky thing: two of the spams refer to your husband. Except, really strange. There's this "X" in the middle. David X. Lev. X, as in marks the spot? Does that mean anything to you?"

***

Nine hundred feet away, in a different pavilion, a man with his name tag and delegation ID turned upside down and in reverse had listened to Lev on one of many designated video monitors, stood up, then gone to a kiosk to buy a quick cup of coffee and step aside, aware of video monitors throughout the halls and corridors. No one was free of the surveillance.

Somewhere in the night, halfway across the world, an old man stooped to gobble ripe boysenberries in the company of… Others.

In an advanced memory ward, across the seas, an old woman remembered back to a terrible time.

Chapter 10
Stalker

At noon, the plenary was over and Lev was met by a long line of old acquaintants and newcomers who wanted to say something, in a few cases grab his autograph on a napkin or tell him that his data was troubling, or simply shake his hand.

Meanwhile, Mal was listening to his messages, mostly from Josh.

"Tell the professor something royally bizarre is going on, Mal. Don't call. It's all fucked up. I talked to his wife. I know, that's how weird. Anyway, better to stay off the phones. Seriously. Something's up. Did you guys tell a bunch o' people, a couple with totally strange accents, where you're staying? Why, outta curiosity?"

"Check your messages," Mal said to Lev, as he finished with the handful of admirers.

Lev did so. It was Sasha. He listened and Mal noticed a distinct change in the professor.

"Time to go," Lev said with a burdened air.

They grabbed the equivalent of Subway sandwiches and Snapple, then sought out the nearest available seats on a golf cart. Lev needed to be at Pavilion 3 by 12:45, and the rest of his day was packed with meetings and speeches he was compelled to attend.

By 7 p.m. he was ready to call it a day. He and Malcolm were walking through Pavilion 3, Mal trailing behind, trying to read his emails without tripping on the journalists' wires, as they were intent upon reaching Pavilion 1, where they knew exactly where to go to board the buses that took delegates back into Rio every half-hour, a long hour of traffic through pelting rain. The bus would stop at a dozen different points, one of which was half a block from their quaint little hotel, three blocks behind the beach.

But before even exiting this third pavilion, Mal had looked up from his devices just in time to notice something at one of the main media message centers, a long wall of notices pinned for people who, for whatever reason, were not receiving (or bothering to glance at) their text messages. Lev had forgotten to plug in his iPhone, now out of power, so he was in electronic exile. He really didn't care, except he did not want to miss a call back from Sasha.

"You're on that bulletin board," Mal said, reaching for the note amid a miasma of names, in bold red, a folded-up little bit of origami-like messenger paper, a Post-it, firmly pinioned with three yellow thumbtacks on the #6 expanded polystyrene plastic wall, what some might call styrofoam – the generic kind of stuff being reported on as more concentrated, along with plastic-derived nanofibers per upper column of marine layer throughout the oceans than zooplankton.

There were hundreds of such messages; a madding crowd of acronyms and cipher: *Ron at UNHCHR, Heather from UNODC, lunchbags for UNDP see intern, UNEP-Nairobi, back at hotel, nibbles, dress cazh; Glorietta, CBD staff, 11:40 pm at the bar, UNU – Barensons waiting still, 6:15, where are you guys?, WHO, UNAIDS*, etc.

Mal decoupled the wadded-up telegram from the bulletin board and handed the rumpled little square chunk of paper to Lev. The professor stood there, wondering who could have posted it, and why.

"URGENT! please, we must speak. Tonight, if possible? Better in town. Where are you staying? I am techsupport." And a given phone number.

Lev shared it with Malcolm.

"Cool," Mal said matter-of-factly. "Some kind of stalker!"

"Not cool." Lev tore up the paper and tossed the fragments into a nearby recycling bin 10 feet away, a few pieces landing on the floor. As they strode off, without giving it a thought, Mal's eyes caught yet another message.

"Sorry, boss. There you are again."

"Well, fuck this," Lev intoned in hushed frustration, reaching for the second such note bearing his name across the thirty-foot span of board. The same type of yellow thumbtacks, and again his name in red, Dr. Lev, far to the right of the previous message.

The professor took the redundancy more seriously now. His heart started to race as he read the identical message: techsupport.

"What is this shit?" he carped.

Chapter 11
Techsupport

Mal reckoned it was an intern who'd been shooting him some data at the last minute from UCLA.

"Or how about a local prostitution ring? I mean, you gotta admit, thousands of geezers everywhere with nothing better to do… sorry, Prof."

"You could be right. Or what about you? Hey, you're unattached, correct?"

"Actually, I've been seeing the love of my life for nearly three weeks."

"Three whole weeks! Amazing! And what does she do?"

"LAPD."

The Professor sneezed. Then sneezed a second time, sloppily rubbing his three-week old hankie all over the middle of his face. "Allergies. Nothing more. Congratulations. A normal person. I need some Sudafed."

"Actually, investigations unit. Her sixth year. She's the real deal. Older woman, no less. Here, I've got some Benadryl."

"Let me have it."

He handed his professor the pink pills. He downed them.

"So what's her name?"

"Militia."

"What's her real name?"

"Melissa."

"Fine. I wish you both all the mazel in the world. It means luck."

"I know. We're engaged."

"I like that. Impulse. Be prepared, kiddo. They all come to the table with lots of baggage. Mine certainly did."

But Lev was more amused than annoyed by the notion of a prostitution ring. He resembled every glint of his age. Deeply veined and freckled hands and clobbered forehead skin; graying wizened hair, reddish and grown long where it could so that he might vainly try and disguise the illimitable truth – all the signs that comport with so much life bound up in one frisson of aged complexities.

Again he tossed the message, making sure it was safely recycled, while also bothering to pick up the couple of scraps from the other message that didn't make it into the hoop previously. He didn't want his name scattered on the floor.

But it wasn't over. Now Lev scanned the enormous board, as thunder and heavy rain were audible atop the large complex, complementing the human stream of ongoing Rio+20 fracas within the vast interior of Pavilion 3.

"There, sonofabitch!" he rasped, a third time that his name propounded into public view: three yellow thumbtacks, and "David Lev" neatly hand-written in red ink. A telephone number beginning with the digits 21.

"Now this is getting freaky," said Mal.

"I don't like it. Not at all." Lev pocketed the message, looked around to see if there was somebody watching him – scanning all sides, suddenly feeling rather naked – and wandered over to the nearest journalist port, where he could plug in his iPhone.

"Follow me," he snapped.

He had his charger in the dilapidated backpack he carried over one shoulder, along with his wallet, passport, and note cards. Malcolm was hauling the heavy documents. Lev never carried the briefcase one might have expected of an older scientist wearing a distinctly vintage-style suit with permanent wrinkles and the odd stain.

His iPhone was now charging where he sat amid a long double row of strangers all sending messages, filing dispatches, reporting, communicating, researching, doing everything one did near the end of a huge international affair that, by 7:20 p.m. on the 22nd had been thought by many to have been a huge success on some levels; many others, the by far greater contingent of cynics, particularly among the press, saw it as business as usual, particularly for the high consumerist nations.

…*techsupport*… Lev kept wondering, certain it was a local Brazilian phone automative messaging service.

"You'll get the phone bill in a month and have a minor shit fit. That's how these international telecoms screw you," Mal reckoned.

"Why three hand-written notes? There's no sense to that!" Lev wondered aloud.

"Watch my phone. I'll be right back."

Lev walked over to a security guard – they were everywhere – and asked her if the message board was under video surveillance. She shook her head that it was not.

Lev didn't believe her, but it was obvious this would not get him anywhere.

Lev returned to the journalists' pool, where Mal sat reading his own e-mails beside the jack in which Lev's iPhone was charging.

"Anything of interest?" Lev asked.

"Something about spams. Josh is concerned. Does 'KS' or the words '*iber azhentsy*' or '*promotej*' mean anything to you?"

Lev thought for a moment. Then exclaimed, at a loss, "No. And who is this Josh?"

"Balfor, remember? He TA'ed for me last semester the week I went to Yale for that conference? Dry forests of New Caledonia."

"Right. Good guy. Spams?"

"He didn't want to go into detail. I'll call him from the hotel. What are you going to do about your new groupie?"

Lev pulled his iPhone out of the jack. It had sufficient charge and he placed the local call. The number rang seven times and he peered anxiously at his phone, prepared to hang up when a Portuguese voice answered.

Chapter 12
The Other Alfred Wallace

"Café Alfred Wallace, *sim*?"

Lev heard a clamor in the background. Clearly a nightclub. The voice sounded like that of a busy bartender, perhaps a bouncer. He hung up.

Bizarre, he thought. Alfred Wallace? Alfred Russel Wallace? One of Lev's heroes. He had frequently regaled his students with tales of Wallace's life. The man who at age 25, with his naturalist friend Henry Walter Bates, left on the seagoing *Mischief* for the mouth of the Amazon in early 1848 and spent four years collecting species, initially insects.

A well-known street near where Lev and Malcolm were staying, in Leblon, was named after Alfred Russel, a well-known Brazilian who did something or other and by coincidence shared Wallace's last name. But that was it. Except for the plaque for Wallace in Manaus, capital of Brazil's State of Amazonas, Lev knew of no streets or parks or the least institutionalized mention of Wallace anywhere in Rio.

"Mal, you didn't happen to see a Café Alfred Wallace anywhere near our hotel?"

"No. As in *the* Wallace?"

"Different Wallace."

"Not likely. Somebody knows you've written about Wallace. That's like totally stalker behavior."

"Stop it. That truly does not set well."

Wallace was a major feature of Lev's doctoral dissertation generations before. Lev had admired the tenacity of Wallace's untutored passions to wrestle with two great ideals: evolution and natural selection.

And to have done so with no trust fund, no wealthy family backing his (then) godless obsessions with many of the harshest terrains on earth. The headlines, especially the Javanese/Darwinian connection – who really first came up with the idea of natural selection? – could not help but jump out at history.

"So what are you going to do? Are we leaving or what?"

Lev picked up the phone and impatiently re-dialed the number one last time. He had a nagging sensation. This could not have been an accident. Someone wanted to speak with Professor Lev, calling him David.

The same voice answered. "*Sim? O que posso fazer para você?*"

He didn't speak more than a word of Portuguese. "This is David Lev. I was given this number."

There was a pause, the speaker moving between languages, then uttered in a voice of panic: "*Espere um momento*. Moment, please."

Chapter 13
Probability Errors in the Rhombicuboctahedron

Four months, 9 days earlier. A nervous man – one might think, either at the end or the beginning of his tether, an idiom that in his area of the planet actually meant rope – was noting in his little research pad, with a #2 pencil from habit: 53°55'53.25"N 27°38'45.72"E, 116 Nezavisimosti Avenue −20 degree Celsius. A pleasant enough evening.

The streets of Minsk were salty causeways of white frost on black ice, laid up against the impressive white Siberias of parkland beneath deep snow on all sides of the National Library, with frozen fountains reflecting the official, highly touted 140,000 tons of *bibliyateka* – a mass 14 times that of the Eiffel Tower. With two transformers and nearly 115,000 m^2 of librarian dreams shimmering in a continuous light show, like a 1960s rock concert from 4,500 light sources.

Famed architects Viktor Kramarenko and Mikhail Vinogradov had designed the spectacular and forbidding diamond shape – a rhombicuboctahedron, the crown jewel of Belarus: 18 squares and 8 triangles oriented on a podium enshrouded in mirrored glass – that any student of library architecture could easily access – and an interior thermos flask, as noted in the energy literature, to keep its millions of manuscripts, and visiting researchers, cozy and warm on a night like this.

Among those intrepid interlopers was a circumspect man in his early 50s with a nondescript shoulder bag, no computer, not even a cell phone, searching for something through the labyrinth, with its heavy layers of video surveillance.

He was but one lonely bibliophile out of about 18,000 researchers associated with the Belarusian National Academy of Sciences, who did most of his work across town at the Yakub Kolas Central Scientific Library, when he was not hundreds of kilometers away, often on his hands and knees in Lyme disease-infested swamps.

For a mycologist to find his way towards very obscure genealogical information in this 23-story, 24-sided glittering "rhombi" was dangerous and, improbably, a bit like turning up a new species of ghost orchid, the last of which had been discovered sixteen years before.

But Taman, named after that "most miserable dump of all the seaboard towns in Russia," as Lermontov had written of it in *A Hero of Our Time*, was determined.

His father, Isaac, had chosen the name Taman, thinking that all of history would prove to be ironic, the stuff of heroism during turbulent times. Belarus had endured her share.

They had their own exquisite language, although most spoke Russian. Russians, however, found Belarusian an exacting, if unlikely linguistic muddle. Most didn't even try.

Isaac and Anna Chernichevsky's only son, Taman, mirrored all of that fragile political insanity. Isaac's father – Taman's grandfather, Nicholai (alleged to have been a monk), had fled his homeland to the south; his Uncle Dmitry to the north; his cousins Ossip, Svetlana, and Hanna, to the east and the west. Another one, Petr, had been executed at the end of the Stalinist regime, or so Taman had heard as a child; Vasil and Sergiusz had been shot by the Nazis.

Still others had died of TB, or pneumonia, or, if they were lucky, had been assimilated into the USSR after incredible tales of long marches, hiding out, sleeping in barns, or under bushes to protect themselves from the sheer panic that frozen stars could induce during a starved and ruinous exile.

Some, possibly many, were anonymously memorialized at the so-called "Graveyard of Villages," a tribute of entablatures at one of the 185 known towns across Belarus entirely torched by the Nazis during 1943.

At least five relatives that Taman knew of, including his grandfather and countless distant cousins, were most definitely Jewish – descended from a Jewish mother. He was quietly but resolutely proud of that fact. Some had been caught up in the Polish/Ukrainian Visla operation and another in the Potsdam Convention tumult. Several other relatives were half-Catholic, but not practicing.

Taman's father had once described himself as a "mystical atheist" of necessity, given all his family's ordeals and his own melancholic personal past of dreams unfulfilled – routinely signing off on ever-insufficient government food staples (even during at least one well-remembered famine) in an office for over twenty years in Minsk, a Kafkaesque civil servant until the "day of independence" (August 25, 1991) when he chose to retire, but not without selling (illegally imported) Studebakers on the side, out of a dank, abandoned munitions dump, for extra rubles.

As for Taman, with the brain speed of an Android outfitted in grungy hiking gear; a whiz kid of the new free Internet generation, he was all science and the spiritual love of forests. And as for those Dark Ages vicissitudes, they'd been visited upon most Eastern European families for well over 75 years, certainly among the Chernichevksys.

And on this night, thought Taman, carefully studying certain documents translated from Hebrew into Russian, it might all start coming home to roost.

Chapter 14
What Would Truman Have Done?

John Vespers paid the taxi driver and ducked into a noisy hamburger joint in Georgetown, where summer vacation was in full swing between tourists and students, all of whom seemed to have mobbed the otherwise unappealing restaurant known for its three pounder. And once, in an earlier incarnation, was said to have been the favorite hang-out for Soviet spies in D.C. They loved American hamburgers above all else.

Two hours before, Vespers had wrapped up a meeting at the Harry Truman Building, or, as most tourists and locals knew it, the arcane "Foggy Bottom," where Vespers's office was located in a distant corner of the proletariate-like third floor at the Department of State. He had been briefed by several of his underlings in CBA (Commercial and Business Affairs), a division of the Under Secretary for Economic Growth, Energy, and the Environment, which he co-headed, and none of it good news.

"We're totally locked out," Pelmar Setting, his science advisor, informed Vespers. "You'll never get anyone in there. Of course we have eyes – erratically – above, but none on the ground that we can trust."

"What about that girl? There was some girl?"

One didn't dare hold back from Vespers. He had a very impressive laser stare that suggested he could burn through a telephone book. Setting had glanced at his own assistant, Cassie, who showed extreme doubt.

"So what would Truman have done?" Vespers asked, lobbing the proverbially loaded bomb.

"I don't follow, sir."

Vespers wondered aloud, "Obviously we have people up in Minsk?"

"Yes, but that's… that's day and night, John."

His iPhone was buzzing – his personal iPhone, not his Department of State iPhone. Caller ID unknown.

"Yep?"

"John, it's Sarah. Your boy has definitely left the reservation, big time."

"I expected as much."

Sarah Jespard, his number two and a graduate in international affairs from Sarah Lawrence, replied after a brief silence, "What do you want me to do?"

Chapter 15
A Room with a View

Claire Upton had opened up a sizable file on the entire National Park, both sides of the border, but particularly that of Belarus. With diminishing storm fervor in her area of specific concern, her "room with a view" had conceded more and more GIS resolution on what had emerged as some kind of Level-B Operation, one worthy of her department's interest.

"Jake, you gotta look at this," she said, buzzing her boyfriend down the hall.

He came to her office immediately.

"Are you kidding me?" he said, examining the replay in slow-mo at 3-meter resolution.

"What, some kind of hunter's night out? Turf warfare?"

They both watched as the shadow figure stepped out from beneath an enormous tree – perhaps a spruce, or an alder, or maybe an elm – pulled a metallic crossbow at full extension, waited five seconds while the target moved closer, and then unleashed the full force of the arrow that flew at a phenomenal speed and went in through the chest and out the back of the victim – another hunter, by the looks of it, possibly military.

Then the figure deftly placed a second arrow in the heavy-duty contraption, took one second to aim, and sent a second victim, also at well over a two hundred yard distance, flying off the ground – this time, the arrow pinning the man to a tree, through one of his eyes.

"Lovely," Jake said with near stoic disregard. "You all right?" he asked as he stroked Claire's hair.

"I know nothing about archery, Jake, but I looked this stuff up. You can buy it online from half a dozen sporting goods stores in Eastern Europe. Those arrows were traveling at over four hundred feet per second."

"It's really hideous. What's going on?"

"Don't know. We need a reliable back channel. Your friend who likes glazed twists?"

"You're hungry. You had a huge breakfast."

"Jake," she said as she took his hand from her hair and touch it against her belly.

"No way!" he half-gasped.

Chapter 16
In and Out of Shadows

At a stoplight, Jake turned and said, "When did you find out?"

"At my physical, last week. I was already pretty certain."

"Our parents'll freak out."

"Not mine. And I'm sure your mom will be cool with it. She knows from experience."

"What are you saying?"

"There are some things, Jake, that mothers don't tell their sons."

"My dad is military through and through. You know how he went apeshit when I grew my hair long and got an earring."

"And you prevailed. You said he adores me."

"I said he approves of you. Fine. So he worked in the Pentagon for thirty years and can't shake certain genes that come with the territory."

"It's our lives. I'm not the least worried and neither will your mom be. So zip it."

Jake Cosgrove and Claire Upton. now eight months away from a long-planned joint holiday in Paris – they were going to be married there, beneath the redwood near the Hôtel de Crillon – strolled from the parking lot into a sushi den at Tyson's Galleria, were seated in the back, near the restrooms, and promptly placed their orders.

"So what do you want me to ask him?" Jake said, changing the topic. He was glowing with the news as the waitress took their order.

"Here's the weird part. It's the first time in a year that I've seen so many visitors to the Belarus side of the National Park – like eighteen. Got all the license plates, none of which helped. Rental cars, or official Belarus government vehicles, aside from a few private citizens. One embassy car with, it would appear, stereotypical blond, Aryan guests, got stuck forty miles away in deep snow last year."

"Military?"

"Bad ground Intel, clearly. They never even thought to call in advance to ask about the roads at the closest pit stop. No AccuWeather, and no Triple A."

"Cute."

"Aside from a couple of scientific types – one of whom is basically living there, I suspect a professor attached to the country's National Science Academy, apparently a couple of grad students, a research assistant or two from the looks of it – I've tracked in and out of shadows a lot of movement – the others were consulate or international trade relations departments. And so here's the question: What can your friend find out? GCHQ's got far more traction in Minsk than the U.S., obviously."

"What does your gut tell you?"

Their saki was served. Jake poured, and repeated by way of a toast, "To Junior!"

"Jake, there are two."

He was stunned.

"Are you happy with that?"

"I'm speechless. Ecstatic. Do we know what they are?"

"I suspect they're humans, Jake," she smiled, sipping her saki.

Jake gave her a glowing look.

"I don't need to know."

"I'm due for a raise. "He again lifted his saké glass, "To you, Mrs. Cosgrove!"

Holding up her own glass, she got it out, "Actually, I've given it a lot of thought. I prefer Ms. Claire Cosgrove Upton." She giggled naughtily but with resolve.

"No, no no … don't do that."

"You pussy," she exclaimed, then put her free hand over her own lips.

The alcohol had gone straight to Claire's head quite quickly. "This is my last drink, until delivery, just so you know."

"Should you even … "

"This one is fine by me. I won't tell."

"We're gonna have two kids. I'm speechless!"

"That would be a first."

"So what do I ask?"

"Just tell him there's a feeding frenzy going on in that national park – and you don't mean California rolls – and that your fiancé … "

"He's seen your picture."

"Did you tell him we're engaged?"

"Not yet. Didn't want to jinx it."

Claire kicked him under the table then said, "Tell him I need any intel he may have. And then see if he's in any way tracking possible missives, diplomatic pouches, anything from our friend Henry's office about what's going on there."

"Be right back."

Jake went into the men's room, locked the door and sat on the toilet, surrounded by thick brick walls, pulled out his ancient G1 T-Mobile cell, with its QWERTY keyboard, perfect for situations like this; he peed, then placed a call to his former classmate from Columbia University, Edmund (Eddie) Ellis Rosenthal. He'd gone home to London after graduating nearly at the top of his class in mathematics and now worked in The Doughnut. He had relatives from the States – hence, his middle name – and he and Jake had sustained virtually parallel careers. Same age, equally adept in their identical fields, just different nationalities.

The kind of long-lasting friendship that typically could get dangerous.

Chapter 17
Fungus

"Give me twenty minutes, Jake. Should I call you on this number?"

"Yes. Thanks, Eddie."

Jake resumed his lunch with Claire. "He'll call me back shortly," he said.

It took no more than fifteen minutes. Jake held on to his cell, headed back into the bathroom.

"Anything?"

"I checked a few key sectors, the ones I can informally hit up. Nothing certain, but it does seem that there is a lot of pharmaceutical interest in that forest. Nothing new, mind you. Especially on the western side. But this is the first time –I have it on some authority – that there has been such scrimmage on the eastern side. One of them is a major European big pharma with roots in Minsk and Moscow. Another, their National Academy of Sciences. My hunch, proprietary revenue sharing in something."

"In what?"

"Truffles, who knows."

"That's weird."

"Whatever."

"Eddie, one more question."

"Uh-huh?"

"Do you know Henry?"

"Henry … Henry. I know lots of people named Henry. Loved Henry the Fourth, both Parts One and Two."

"Any who hang out like, say, in the Truman Building?"

"Ahh. No. Don't know that one. But give me a few days. Say, you still getting out for any rugby?"

"I'm getting married, Eddie. To a fox. Fuck rugby."

"Congratulations, Dr. Cosgrove."

"Children on the way, Eddie."

"Double congrats. By the way, be careful."

"What? You have a son. Has it been so bad?"

"Not that. There was one other thing I uncovered."

"Shoot."

"Some scientist in Minsk has been poring over Jewish genealogical records at Yad Vashem. The guy works in that National Park, or so it appears. And from his web browsing footprint, he's definitely naïve when it comes to security issues."

Chapter 18
An Uncommon Urgency

Taman's mother, Anna, still played the viola da gamba – what an intoxicating sound it produced in her hands, he thought – and she had once been a formidable interpreter of Max Bruch and Brahms. Taman would frequently fall asleep as a child to his mother playing her rendition of Richard Strauss's "Don Quixote" viola parts, as well as Edward Elgar's "Enigma Variations." He felt blessed to forever carry those hauntingly beautiful sounds within him.

Now, she spent her time worrying about her husband, who had survived two strokes – reading Russian translations of another stroke victim, the Swedish poet Tomas Transtromer, to him at home – and occasionally giving music lessons to Taman's wife, Sofia, and his two boys, who were already scouting around for universities. And as often as possible she would visit her Aunt Sarah, who at 81 had been diagnosed with mid-range Alzheimer's – whatever that really meant, he wasn't clear – and had been languishing for three years now. It was all so vague, thought Taman the scientist, who had his own particular interest in Sarah.

There she was, in a kind of pure delirium down in Brest, Alzheimer's, or just old? She lived in a memory care unit that was immaculate, smelling – not of urine – as so many people had remarked about such melancholy places – but, rather, of poppy seed pastries and honey. Moreover, the nursing home was located very close to where Sarah had grown up, just across the border in Poland.

A kind of geographical comfort only those who actually understand it can attest to. Knowing that just out the window... over there... across those fields...

Sarah somehow recognized the proximity of Poland a mere dozen kilometers from her bedside – the languages and dialects enunciated in her penumbra of dreams, gestures, and daily food service. The mushrooms. As well as the TV stations she watched, without actually watching. That was the disease eating at her brain.

Taman had gone to see his Great Aunt (Ciotačka, Auntie) a month before. It was a matter of uncommon urgency.

Sarah thought he was her older brother. She was yelling that morning about fat people stealing into her room, coming in through the window from the shopping mall; a double-locked window permitted a view over a frozen landscape of renovation and re-paving, between Budyonny and Dzerzhynsky Streets.

"Sarah, it's your room. Don't worry. I'll get rid of those obese farts," Taman said authoritatively, speaking in a composite of Polish, Russian and Belarusian.

"And they're ugly!" she added, and broke out in a naughty laugh. She had only recently been confined to a wheelchair.

"Most certainly," Taman concurred with a conspiratorial grin that was nothing less than a family signature.

Sarah slapped her thigh and they went and sat together on her bed. It was a king-sized hospital bed with railing flaps that were now lowered on account of a recent incident. The morning nurse had found her on the floor, a gash on her forehead. But Sarah picked at stitches and butterfly bandages. They had to leave the wound cauterized. It bulged purple but would heal.

"She's not in any pain," the nurse reassured him.

"Don't be ridiculous. Of course she is."

Taman knew the routine. It had only escalated. Palliative care. All the glib denials by everyone in the process, before the inevitable moment. But not on this day.

Chapter 19
Eyes on the Ground, Spring 2012

Sarah phoned JV – as she referred to her boss, John Vespers – on a landline from her lawyer's office in Alexandria. She'd had an appointment for weeks to go over a few details of her recently deceased mother's estate.

Vespers picked up. "Who is this?" He saw "Esquire" as part of the ID.

"JV, it's Sarah."

"Where are you?"

"Personal. Not relevant. Anyway, I'll be brief. Checked the master list, everything. There is that young woman – the one we've discussed, that girl who got a job as an assistant at the same research station in that place."

"Yep."

"Then you probably remember she's the same one we tried to contact. She had unique access for nearly a year to Lukashenko's private meetings, itinerary –"

"How'd she do that? You're talking the ultimate control freak."

"Her first cousin worked in his executive offices. She fed dozens of leaks under cover of some Ukrainian Free Belarus journalists' site."

"Can't trust her."

"Not sure about that. Eventually, she felt the rising temperature and promptly removed herself to some innocuous position inside a science division of the government. She has a masters degree in forestry from Moscow University and got hired by the professor running the show out there in the woods. I think she could be our eyes on the ground, if we can find a way to get to her. I don't know about her boyfriend."

JV was just on his way back to the Truman Building to speak with foreign nationals who'd flown in that morning about some specific commercial prospects in play that needed refining prior to the upcoming UN Summit in Rio. He had concerns about one of the Mission's proposed list of delegates. He was equally concerned about the amount of internal traffic being generated by this "project."

Moreover, the Polish point person was no longer among the living.

But now, his own mission just got beautifully amplified, he deduced, spinning out scenarios in his mind.

Chapter 20
The Road to Utopia

Taman had come bearing brownies, and with a dainty voracity she consumed two whole ones while he put on one of her favorite movies starring Bing Crosby. The DVD player on her private television was well used. She'd watched "Road to Utopia" at least dozens of times. It was Alaska, the Skagway in particular, that was the lure – a familiar enough place, to her – and the gold mine, of course.

"We went out several times, you know," she said, referring to Crosby.

"That's what you told me."

"It's true. Both he and Bob Hope wanted to marry me."

"What happened?"

"I had to choose. They both could sing, you know."

"How could any woman in her right mind have been expected to make such a choice, especially in the middle of winter in Alaska?"

"Simon promised we'd sail away. Any girl would have gone to Tahiti," she said dreamily.

"Simon who?"

"Exactly," she said, without explication. She hadn't explained much for several years. Although, admittedly, Taman had only come visiting as of late.

"And can you believe it? They were both married to other women!" she let ring, with a giggling, dreamy déjà vu.

"That's Hollywood for you."

"Of course. I lived there for years, you know."

"How was Hollywood, Auntie?"

But she just chuckled, the question having nowhere to land in her mind, or so it appeared to Taman. This was the condition. What was possible, for her, lay in the unknown zones.

A new nurse popped in. They had the annoying habit, Sarah indicated, of changing constantly. She could never keep track of all her new friends.

"Hello Sarah! And how are we doing this morning?"

"Is it morning? Well I'll be damned. You've met my older brother?" She couldn't remember Taman's name just then, as she looked cautiously under the bed for her dentures. She'd lost them again.

"Of course. Oh ... I see you brought Sarah some treats? That's her favorite. Well just let me know when you want your lunch, dear."

"What's for lunch?" Sarah declared imperially. "They all call me 'dear,'" she told Taman, indicating that she quite easily saw through the entreaties.

"*Machanka* and your favorite cold sorrel soup. We also have fresh dumplings today."

"Just make sure no more fat people! Not in my room!" she hollered at the nurse.

"Absolutely, dear."

"And don't call me 'dear,' damnit! By the way, whose clothes am I wearing?"

The nurse was made of Teflon. "I don't know – yours; and how beautiful they are on you. Perfect fit!", my dear.

"Fat people! Get rid of them!" Sarah shrieked.

"Road to Utopia," the nurse acknowledged. "Sarah knows every word of it."

"Of course I do. I almost married both of them, you know."

"Yes. And boy, are they the losers."

Sarah chuckled. "You bet they are!"

She found her dentures and turned for privacy to put them back in her mouth.

"I'll close the door," the nurse said.

Chapter 21
The Day of the Chicken

"Thanks," Taman said. Then, holding Sarah's hand, "Your young friend, when you were a little girl? You said the Germans had come into the village and made him shoot at a chicken?"

"Chicken? I don't want to talk about it."

"Auntie, please, you know I love chickens."

"Well so do I!"

"I know that. So listen. Last month I visited you and you told me a story about a boy and a chicken."

"Uh-huh."

Taman was resolved to unlock this riddle he knew meant everything.

"A chicken. The Nazis. Remember?"

At first she stayed glued to the little plate with the crumbs of remaining brownies, dabbing her finger on it to pick up every last morsel. Then she glanced up to see the movie for the 101st time, one would think.

"You said something about a particular event, a horrible story, as I recall."

At once: "Those fuckers," she barked defiantly. "You know we got every last one of them."

"Didn't we, though!"

She clutched his hand. "Marshall! And Roosevelt. And then the moon!"

"You were telling me about it," Taman urged her.

"Terrible. It was terrible. They just stood around laughing, drunken pimps. Imagine putting a machine gun in the hands of a child."

"You said the bullet backfired. What did you mean?"

"I don't know."

"Auntie, tell me what happened with that chicken?"

"Oh, there were lots of bullets. You know how a machine gun works, surely? Fuckers, every one of them be damned!"

"Yes." She loved "damned" and "fuck."

"Well, then. It was some kind of machine gun. Oh my … !"

And wham! She was there. Could not remember a phone number of two minutes ago, or even who Taman was, exactly. But she could remember everything from more than 75 years ago, down to the color of the tile on an adjoining window...The face of a goat down the road...The light pouring through the stormy clouds that day. All the faces gazing off, blank stares towards the infinity of a darkness that was closing in on them, as fast as a quickly moving thundershower.

Gazing right at the instant it happened. "The bullets fired and hit metal. That lovely little chicken escaped. I think the bullets went into your uncle's roof. I mean for Christ's sake, the boy was my age. I might have married him someday. What a handsome man he must have become."

"Become? What did he become?"

"Who?"

"Sarah, your friend, the little boy"

"It's a secret," she said, and started to cry. "Anyway, I wrote it all down, everything."

"Where?" Taman had been hoping for this.

"I told you: It's a secret."

"If anybody can keep a secret, I'm your man."

"Promise?"

"Promise."

"Well, it's all in my diary."

Now Taman could hear the blood loudly coursing through his veins. "And where is it, Sarah?"

But she couldn't staunch the flow of tears, and shook her head. She couldn't remember.

Taman stroked her long white hair. There was a beauty salon in the nursing home and she went once a week: hair, nails, toes. A new shower with a very modern granite bench, fresh from the mountains of Poland. Every day – no baths, she couldn't get in or out of a bathtub.

It was now fixed in her mind, that revolutionary 48-hour event she would never forget, and had never told to anyone before Taman.

Suddenly, "You see, he couldn't aim a machine gun. He could hardly lift it. And his hands were trembling. They were going to execute them all, you know."

"What happened on that terrible day, Auntie?"

Her expressions were gasping along a gauntlet that Taman was desperate to understand.

Then, when least expected, a grin materialized upon Sarah's ruddy cheeks, and her eyes blossomed: "One of the bullets fired back and hit the German bastard in the gut. He fell over screaming, but he wasn't dead. Taught them never to screw with a Jew."

Taman nodded. "I like that."

Sarah giggled in her manner, off somewhere.

"Of course, your grandfather was the one they made stop the bleeding. He was the only doctor in town."

"He was my great-uncle, not my grandfather." But for a brief moment, Taman was off his guard, trying to make certain of a murky genealogy.

"It was bad. The second bullet hit Simon."

"Simon?"

"Yes."

"What was his last name?"

"The chicken got away."

Chapter 22
Deep in the Forest

The archer literally ripped the dead man from the tree, pulling off a messy section of his brain to do so. The scalp was flayed. Then, digging furiously, he buried him and the other man in a deeply concealed cache of rotting logs, under seven feet of snow some 900 feet away from the site where he had murdered them. The earthy ensemble was primordial, and with a slow thaw the decomposition of the bodies would be rapid, he knew. The odor wouldn't last.

He was careful to bury their rifles with them and to walk two full days just over the border into Poland to ensure a chaos when the dogs were unleashed, which they surely would be.

Then, skirting key positions, staying close to the shadows the way most prey species do, he eventually returned, after four days out, to the research station, careful to have stashed his instruments of death in a place a quarter mile from their camp.

He made sure his appearance now gave nothing of his activities away, and he strolled up to the center, his field notes and pencil in hand, greeted two of his colleagues, and went to his room.

"What are you guys eating?" he asked.

"Fettuccini. Ulyana made it. There's plenty. Also freshly baked bread, just out of the microwave."

"Where is she?"

"Who knows? Probably shooting BBs at icicles."

"Shooting with what? What are you talking about?"

"Lighten up."

At that moment Ulyana walked in. "I heard that!" she scowled, implying no nonsense.

Petrovsky eyed her with caution, as he always did, but smiled with so normal a demeanor as to relegate the least possibility of suspicions to the netherworld. She hadn't a clue. He was certain of that.

"When are you guys going back out?" Petrovsky asked.

"Soon as the prof returns."

"Didn't he just finish gathering up the data from, what, transect 203 or something?" Ivan Petrovsky knew exactly where he was, but it had not served his conflicted interests for many months to share anything with his two colleagues, Yuri and Vladislav.

Both were nearing the completion of their dissertations under Dr. Chernichevksy, like Ivan himself. Yuri would no doubt go back to Warsaw where his family lived and look for academic employment at a Polish university. Vladislav was hoping to secure a job in the Belarusian Academy of Sciences that would enable him to continue research in the field, an unlikely prospect given the economic downturns. Neither seemed in the least suspicious, just focused on their research, whereas Petrovsky was certain that something was very wrong with Ulyana, as well as her boyfriend, Jakob.

Petrovsky knew she had a gun somewhere, and not one for BBs, deducing that fact from a metal fragment he'd bitten down on in a rabbit stew she'd served the boys several weeks before. He kept silent about it. He just couldn't figure out where she kept it or how she'd obtained it, unless she had military ties.

When his research grant was not renewed, he chose the option that had drawn him in: a young, very attractive American exchange student he met during *chas peak* – rush hour – on the Moscow subway. She said her name was Sally, and that she was a college student from the East Coast. But he quickly found out the truth: she worked for some intelligence agency in the States. Knowing that, Ivan nonetheless agreed to provide information about mushrooms that any grad student with a knowledge of online research papers could more or less cotton together, in exchange for the huge sum of US $3,000, wired unsystematically, usually arriving within 45 days or less, to his widowed mother's branch of Credit Bank of Moscow in a far northern suburb of the capital, alternating with his grandmother's or sister's accounts at the same bank branch.

Ivan's older brother had been one of the unfortunate victims of the 2002 Nord-Ost siege of the movie theater in Dubrovka. Some 850 hostages had been taken by the Chechen terrorists. More than 130 of them died from the poison gas the Russian anti-terrorist teams fed into the ventilator system. Nobody, said Sally, would ever question reparations coming to the bereaved mother, sent – by all appearances – from the Reparations Bureau, a dummy division of the Russian Federation.

Now, those days were long gone. Three weeks after Ivan's visit to the Church Bells near Red Square, both his mother and grandmother died within 36 hours of each other. Ivan's mom had been visiting his grandmother in Moscow's Sergiev Posad area, famous for its enormous medieval monastery, the Trinity Lavra, and also the birthplace of Russia's greatest painter, Andrey Rublev. Ivan's *babushka* had lived there since first moving from White Russia to Moscow with her husband, long gone. On weekends, Ivan's mother usually took the train out to the northern suburb to be with her mother, visit holy places, and have a meal. On this particular Sunday, as was the ritual before Ivan's mom took the train back to her neighborhood closer to Moscow city center, the grandmother fixed a lovely repast.

Both women were rushed to the ER, and both died from a mysterious poison that turned up in the teapot from which their servings originated. No culprit; horrible deaths. Faces turned blue. Ivan had no doubt about what had happened, possibly the seemingly innocuous pretty little yellow flower, *Gelsemium elegans*, lethal if properly prepared. It had been used in several known Russian assassinations.

He still had the handwritten note from the American giving him emergency medical numbers in greater Moscow. It all made sense.

And, not surprisingly, no more money ever arrived.

By that time, Ivan's loyalties were undivided: he had always greatly admired his teacher in the forest, even in his undergrad days. He knew Taman to be noble and selfless. Now he realized that the whole lab, and its core activities, were being scrutinized – not that he could fully sniff out the level of guile.

With such weather, the worst he'd ever seen, satellites weren't supposed to work, he assumed. And he had only seen one drone in three years of being out there – a UAV sending signals on the transboundary wisent migrations back to the Polish National Park system, nothing sinister. That's what he'd heard.

Yet, since even before the time of his mom's and grandma's murders, he knew precisely who it had to be. The stakes just erupted with the murders, and he couldn't possibly predict the level of international ties that would drive people to commit such heinous acts.

So in addition to biochemistry, he endeavored to come up with a plan to master other skills, at great risk to himself. Those of an assassin. He'd grown up hunting with his father. He despised everything about it, having turned his heart and mind thoroughly towards nature. Yet, those skills had already been put to good practice. Vengeance could be a beautiful thing, he discovered. And, having taken money, he felt the full brunt of guilt, of family blood on his hands. Now, a potent force flowed through his veins towards redemption. Whatever it would take.

Moreover, it was also clear to him – but apparently not to the others at the research camp – that there was an actual insider. It could only be Ulyana or her boy friend, Jakob – his was a mysterious job, working in the vodka trade, or so he said – and he'd join Ulyana every third weekend or so – about five hours by train, on average, from Minsk, plus the bumpy land journey of another hour or so – and they'd fuck themselves crazy, like jackrabbits.

© M.C. Tobias

Chapter 23
Simon Stylites

"What about the boy, Simon?"

She stared with the vacancy of true terror gone blank: "The bullet hit him in the face. Blinded him in one eye. And you worry about his last name?"

"Which eye, Sarah?"

"If it weren't for that day, I would never have married Crosby and Hope."

"Which eye?"

"Where's that stew? And those dumplings? Please, go get the nurse."

Taman went to the door and slipped out to the nurses' station.

"Lunch?"

"Five minutes."

He walked back to his aunt's room, # 29.

Once beside Sarah on the bed, he took out his Android and slid his finger to the precise image in question, taken several months before. Then he drew his fingers to either side, enlarging the image, righting it horizontally until it was the maximum digital strength.

"Does this face – with the bad eye – look at all familiar to you?" he asked.

"Weren't they a riot!" she laughed, fending off the picture on Taman's Android, trying to work the clicker to speed up the movie she had seen endless times, but could not make work.

"Sarah, please. Do me a favor. Just look at this for a minute."

"All right."

She needed a magnifying glass. "Some old man in the forest."

"Do you recognize him?"

"Why would I recognize him? Is that Alaska?"

"Sarah, see his eye?"

"Left eye," she said. "But..." She fumbled with his Android in her hand and studied it.

"No. That's impossible." Her expression evidenced clear distress.

Agitated, she tossed the Android and it hit the wall. Then she raised her gaze, grabbed hold of the clicker and tried to change the channel on the TV. Taman got up and removed the DVD, as the nurse brought lunch on a tray, with some fresh sunflowers, as he retrieved his electronic device from the floor.

Chapter 24
Yakub Kolas Central Scientific Library

15 Surganov Street, Minsk. It was early afternoon and Taman sat in a private cubicle, his occasional haunt, reading and re-reading Presidential Decree #59 of February 9, 2012, issued just weeks before and invoking a "conversion" clause that would demonstrably affect the "Berezinski Biosphere Reserves and National Parks," "Braslavsky Lakes," "Pripyat," "Narochansky" and, most importantly, "Belovezhskaya Pushcha" – "changing boundaries of the zones and regimes of protection and use of specially protected natural territories."

That was the obscurant's language. Full of 1984-speak, rough not just in translation. The same "big-brother" jargon monopolizing the unlikely revenue streams from the eco-tourism now beginning to come from Chernobyl.

Taman's research within the labyrinthine corridors of the "diamond" – numerous visitations and risky on-line chat-room investigations with Israelis, Poles, Americans, and one Russian family – had yielded the crucial link: Professor David Lev from southern California.

Now the particulars were nothing less than an avalanche of urgencies. Such a decree, from the top brass, potentially catastrophic for Taman's entire raison d'être, as it turned out. The Berezinski Biosphere Reserves, Braslavsky Lakes, Narochansky, and Belovezhskaya Pushcha.

He reflected upon the sinister decree: "Changing boundaries of the zones and regimes of protection…" Hazardous vagaries of change might come in the form of machine gun-toting goons; foresters taking the law into their own hands; more tourists flocking in search of wisents or red squirrels or any of the nine-cavity nesting avians in huge primeval trees, some of those trees bearing celebrated names dating back five, six hundred years. It could mean minerals, as well.

A new road, edge effects for which the most recent data was positively devastating, a set of numeric probabilities expanding the ecological war zone to some 70 meters beyond every road. In the Amazon, it was now a known quantum leap: every road spelled growing disaster inside the forest.

In Taman's world, that also meant debris pits, pre-manufactured construction sites, all manner of fragmentation with word of a new administrative annex, extended military walls of electric fences and sensors, off-road motorized paths, and stray backpackers trying to find interesting shortcuts to the Polish border through the forest.

Two international conferences (admittedly, mostly Eastern European in persuasion) were upcoming as part of the Yakub Kolas calendar. The first, prior to Rio, in Minsk, on "Biologically Active Substances in Plants." He dared not go, but one of his grad students, a Russian, Ivan Petrovsky, could do so. But this conference worried Taman.

It was not the people already listed as going, but the corporations that had remained silent, whom he had no doubt would be following the talks most closely.

The other, post-Rio, on sustainable forestry, in Belarus's second largest city, Gomel, the second week of October. But that would be fruitless – opposite end of the country. And it would be freezing cold.

And he needed to act much sooner than October. Things were going downhill fast, conditions verging on morbidity – in the literal sense.

But it got even worse: There were corporate players from Kiev to Brussels to Manhattan, or so he'd translated the rumor mill in his profession. Everyone wanted a share of this suspected windfall.

Whereas Taman saw it rather differently. He had merely a new biological series of variables, the equivalent of a new biome, that shone a totally unexpected spotlight on new species connections, nutrient turnover, leaf litter composition, and old forest in a manner nobody had expected. Like the stock markets – predatory, working by the second, timing quarterly reports and IPOs – Taman's e-mails had been countermanded; a modest five-page article in the peer-reviewed *Nature*, and some local (Belarusian) press, had driven home a message of easily distorted superlatives.

Suddenly, there was no one with whom he worked that he could trust. Not even his own graduate students. He changed his work habits, schedules, never showed up when expected, locked his microscope in a cabinet at night.

Nonetheless, he sent Petrovsky to Moscow on his behalf for the conference – mostly because Ivan was the one person he still felt certain about – and because he was Russian and there would be no travel issues at the last minute.

Chapter 25
The Stampede for a Catalytic Agent

Taman took careful note of each of the 32 regulations stipulated relevant to the decree that would serve as a template for the inevitability of the commercialization of scientific discoveries, to make science pay for itself. Mycologists like him might get lucky and find a new fungal catalytic agent – indeed, he had a long-term experiment in the works that all but promised something rare, valuable, and unheard-of, with a proliferation of commercial applications and a real-time subject about which he literally got migraines just contemplating; the implications were simply too startling.

Adding to the lava flow of nefarious complicities, he knew that what he had stumbled upon was likely the centerpiece of what he and some close colleagues had perceived to be an intensified radar screen of surveillance by the BNAS (Belarus National Academy of Sciences). BNAS high-level authorities were taking a sudden and inordinately intrusive interest in the cyberseedling R&D, with veritable demands for rapid-drug applications and eco-dynamic commercial prospects. The flippant explanation from his boss: the BNAS had to pay its own way, becoming self-sufficient.

Taman knew that that was not the real story, not in a country dominated by social servants and universities controlled by the state. Darkening conspiracies played out in his convolutions of worry. He feared saying too much about it to his wife; somebody might be listening, or she might inadvertently say something to a friend at a shopping mall, a music class, anywhere. Any time.

At the root of his real terror was the very likely involvement of Russian, Ukrainian, or Polish big pharma investment schemes, involving lots of money, so-called "Big Data," and intellectual property rights transgressions. Buyers in New York, France, and Belgium, representing pharmaceutical companies easily willing to cough up $700 million to rush a drug through multiple phases in the Western markets, were lining up, all in secrecy. He simply imagined that this was occurring, lacking, as yet, any proof.

Indeed, he knew that what he had theorized – although without the necessary peer review, an impossible prospect – might well be the game of the century, a contest additionally complicated by an epicenter, a living subject whom no one would want to acknowledge or talk about. Never had such a situation arisen, to the best of his knowledge.

It was this mystery person who had totally altered Taman's world and sent him, terrorized, into the shining diamond, or "Rhombi," as many called his nation's national library. Now, the stakes kept him up at night, and made withholding the person's whereabouts his number one priority – that, and trying to catch sight of him again.

Let them call him just one more poacher, Taman thought, knowing that out of all those satellites in space – and given his country's fanatical obsession with spying on everybody and everything – this precariousness of big pharma involvement… whatever, whoever this person was, could only result in a full-scale crisis. Taman was no diplomat.

He felt trapped. His personal life was in shambles. He diverted all hopes, went to bed frightened and depressed each night. He awoke frightened and depressed. The more he learned, the worse – far worse – it got. He had thoughts of taking his own life. He simply did not know what to do or how to process these data sets, both instinctive and instructive.

The one thing he could do, and set out to understand, was how he might manipulate the end-user rules of the game.

Chapter 26
Thermus aquaticus

Taman thoroughly examined whatever biochemical precedents there were, even if extrapolations from his own study area were as yet vague.

In the late 1990s, he knew that authorities at Yellowstone National Park had joined forces with the private sector to exploit enzyme biocatalysis, transgenic plants, and new, bioactive, commercially viable products "of special interest" deriving from the nucleic acids in samples of DNA from thermal pools. The bacterium, *Thermus aquaticus* of the *Deinococcus-Thermus* group, had resulted in hundreds of millions of rubles in profits.

Taman was a scientist with no head for business. But this thing he was on to, a combination of new species, novel epigenetic expression like none he had ever heard of, and the possibility of a bonanza, had turned his bogs into a potential Wild West, a gold rush. He was in trouble. The plants and animals would be trampled if the truth got out.

He had neither the heart nor the personality to exploit it in that fashion, nor the empirical proof – at least not quite yet. As the time grew closer to the Rio Summit, his fear of his own graduate students increased. He had more than a few good reasons to be paranoid.

Not a night passed when he did not lie awake thinking about amino acids, polypeptide strands, syncretistic effects, and the moral dilemma confronting him.

Amid this flurry of fears, he imagined a nice retirement for him and Sofia, maybe an apartment in the Virgin Islands, or among the lovely hills of Nice, in the area of Monet's Cimiez Cemetery. Taman had once visited Nice for a three-day scientific conference.

He had absorbed inexplicable comfort from Monet's "Water Lilies," and someone pointed to the hotel up above the cemetery where Monet had lived at the end of his life. That whole scenario – the museum, nearby monastery, the ruins, wonderful local restaurants – in a quiet neighborhood overlooking the Mediterranean, all sat well with Taman, after so many years in freezing snow and biting insects.

He did wonder what a "Mushrooms" by Monet might have looked like.

Taman glanced outside. The day was characteristically sullen and cold. A snowstorm was coming in, flakes streaking the enormous windows swallowing the academic manse – the Yakub Kolas – with its glossy and transfixing windows, and daunting interior of tomes and sweat equity. Anonymous personages engrossed in their solitary odysseys through annals of scientific minutiae stepped silently around him.

But for months, the value of Taman's long-lasting monitoring stations in the Belovezhskaya Pushcha, or forest – a National Park that spanned portions of Poland as well as his own country of Belarus – was of little relevance compared to the "bunny rabbit" that was feasting on his mushrooms. A most peculiar rabbit – like the one Sarah so adored in "Harvey," which she had also watched a hundred times, and where Taman drew the link enabling him to at least put a spin on the mystery – identified in just a couple of images he'd managed with his old trusted Nikon D200, at approximately 6:30 a.m. one morning, from nearly 1,000 feet away.

There would be two other equally vague, but nonetheless telling, encounters.

His camera needed cleaning – it was noisy. There were two scratches on the mirror he had never gotten fixed. The resulting imagery revealed what appeared to be loose hairs, stringy, obstructive, irreparable. Any kid using Photoshop could clean it up, of course, but he had not done so. The "rabbit" had heard the old telephoto just barely grinding in manual mode so as to autofocus.

And then the rabbit was gone. An autumnal phantom behind a cluster of huge lindens and hornbeams, and behind those, gigantic oaks, some of them 600 years old. He had not pursued the bipedal creature.

The first time, Taman had figured it to be the one-in-a-thousand tourists who had wandered by mistake into the explicitly Scientific Reserve portion of the National Park. It was most assuredly a person. Either a lost tourist, or a very determined mushroom hunter. Or, worse still, a focused and resolved poacher with the moxy to defy authorities.

Dressed in camouflage resembling the hair of the wisent, or a wolf's hide. And working for one of any number of possible cartels – in-country, Poland, Russia, Ukraine, possibly China.

Someone with local connections could pull it off – they had for centuries, regardless of who the sentries were, or what nationality. This ancient forest, the last primeval forest in all of Europe, had endured everything. It expatiated every malevolent nuance of a police state; of a national park closed to nearly everyone, surrounded by military who'd been ordered to shoot to kill; and a border crossing with Poland with more surveillance than any other corner of the human world.

Taman, and his graduate students – a Belarusian, a Russian exchange student, and a third of Polish descent – had long known the drill regarding unwanted guests: do nothing, say nothing, risk nothing. Disappear. As it was, their tenancy existed on a fragile terms basis: their project could be cut short at any time. The funding cycle was exhausted.

A decision was soon to be made at the National Academy of Sciences. Taman had not published anything in over two years, since his short essay on the eco-dynamics of Eastern European forest systems, which did encompass some new mushroom-related biology.

His excuses for not publishing more had been accepted on the basis of long-term study requirements, the admirably thorough and original nature of his research. It's what was called "good science."

New DNA technologies requiring comparisons. But a newly appointed director had set an unrealistic and punishing pace for productivity – there had been word of talks with Western commercial prospects – and Taman's "esoteric" mycology was becoming less relevant by the day, even though the very opposite happened to be the case. He was now on a tight leash and his nerves were running thin.

If the mushroom patch he'd been studying for years did, in fact, come through for him, harboring the nutriceutical equivalent of the commercially viable *Ganoderma lucidum*, there could be a commercial boom soon to follow, as throughout much of Korea, Japan, and China. Part of that more than $2 billion + global market. Bracket fungi were one of many new waves in fighting cancer.

Division: Basidiomycota; Class: Agaricomycetes; Order: Polypoales; Family: Ganodermataceae; Genus: *Ganoderma*. Species: *Lucidum*. Taman wasn't prepared to say, but it was one of more than eighty known edible, wild polypore mushroom species of the same genus. But this one... this pellucid, green sheen of a little girl ever so delicately making a rotting log her autumn hearth, she was different.

The entire morphology and DNA sequences he'd run of its sporocarp varied dramatically from other structures. It was nothing like the wildly popular Chinese Ling-zhi, or its allegedly close relatives in America, *G. tsugae*, or hemlock varnish shelf, as it was known to mushroom hunters. And given its intense psychoactive properties, was likely deserving of an entirely new phylogenetic nomenclature.

Taman's own experimental consumption of pieces of this bracket fungus, in soup, dried, in all its phases, had convinced him that it could easily induce unending hallucinogenic responses in humans, death, or total psychic transduction: a relapse of childhood, a journey to *somewhere else*, that never ended. It also appeared to freshen one's complexion, killing many species of facial skin mites, and fantastically accelerating the clotting of blood – down to seconds if rubbed in a powdery form on a wound.

It was also delicious. And that rabbit seemed to know all about it.

© M.C. Tobias

Chapter 27
Alice's Rabbit Hole

It was late spring, by Belarus standards, and Taman had been out for three days, in a white tent. The color seemed to deter *Apoidae* – hundreds of species of bees in the forest, and the *Vespoidae*, several hundred wasp, yellow-jacket and hornet species; and possibly ticks as well. White was additionally useful in winter – good camouflage so as not to disturb the wisent, or any other creatures. He had no stove, just the usual pack filled with granola bars, cold cans of tomato bisque, slices of pumpernickel, and chunky peanut butter.

Nearby was one of many streams that provided what was probably the cleanest creek water in all of Europe – his secret source originating in a series of high-tannin-laden bogs.

Eighteen kilometers to the east, closer to civilization but still well within the inner protected zone, lay Taman's compound, a well-provisioned research station with a real laboratory, where his three graduate students – having spent a week amid the usual spring tumult of storms, snow melt, more storms, the ever-present risk of being killed by a falling tree, busily sampling scores of specimens – had already packed up and gone back to Minsk.

This was Taman's own stealth camping site where he could leave his tent fixed with provisions and a microscope, and it could be safely zipped up and secured with a twenty-dollar Serfas Espresso Key Lock, and left for the winter. Of course, if somebody found it and wanted to get in, a Swiss Army knife and a few seconds working the tent fabric would easily wipe out the entire cache. That had never happened. This was still the epicenter of Eastern European wilderness.

Moreover, wisents, with their incessant grunts, words shared between them, had no reason to sample or trample a tent. Not their style. Their non-aggression and shyness were legendary, at least among those humans with no other purpose than admiration and curiosity.

No food supplies were left, other than a healthy trail of granola all around the tent – his offering to the birds, insects and small mammals of the forest, in addition to three broken-up loaves of uneaten bread with generous spreads of butter.

He had decided to stay on one more week. It was less a hunch than the desire to simply be himself again, in the old forest with its pure darkness at night, the sound of howling winds, and all those large vertebrates – any quadruped larger than approximately 100 kilograms, like the wisent, or the skittish boar. As for wolf sightings, they were so rare as to break his heart. It hadn't always been like that.

And in his many years of research in this National Park – the most important in all of Europe – he had heard very few rifle shots, and never anywhere within twenty kilometers of this portion of woodland. Taman could calculate such distances with astonishing accuracy.

Poachers had good reason to fear the forest, for it was teeming with ticks that carried the deadly TBE, or tick-borne encephalitis, as well as the similar Lyme disease that killed humans indiscriminately, including two of Taman's colleagues, crawling undetected into one's ear, or up the other's anus; or simply resting comfortably on the chest or under an armpit, often undetected for many days, even weeks. And then it was too late for the host.

Severe climate change had apparently excited the virus. Data from the Czech Republic, from Sweden, and from Poland had confirmed two-, three-, or even four-fold escalations of the acute, or sub-acute transmissions to humans. Tourism offices were saying nothing.

Any poacher, on the other hand, might think twice about coming into the forest. As for this pooka, Harvey, he didn't seem the least concerned. He did not stand 6'3" tall, as in the movie with Jimmy Stewart. And he was certainly no White Rabbit out of Lewis Carroll. This was a man. An old man. Taman had sworn to an inner secrecy about the event, the images of which he studied intently in his tent that night.

He knew he had made a discovery of devastating portent.

Who are you? Taman perplexed, lying awake in his sleeping bag beneath a stormy sky his last night before breaking up camp and heading home. He was peering through the meager, grainy images he had managed to capture before the subject had disappeared. Enlarging every single aspect of every frame.

This man was no hunter, and no scientist. More like a Russian Orthodox monk/survivalist on an ascetic end-of-life sojourn.

Freckles. Vestiges of red hair...long and shaggy. A beard. Maybe not so old, simply the product of intense weathering. Slightly hunched but sturdy, to be sure. Vibrant, in fact. But his face, or, more precisely, his left eye. It was sunken in. A distorted hollow, most likely the remains of something very bad that had happened. Blind in one eye. No doubt about it. The man's age, impossible to fathom, but certainly in his 70s, maybe 80s. His clothing – vintage, like some traditional player in a troupe, ragged, swatches of fur or leather partially eaten through, as if the same age as the man himself.

Definitely the attire of a caveman well conditioned to the perils and obstacles of the landscape. He must have helped himself to a carcass, using a hunting knife, presumably, to shave the clean skin with its deep fur, probably from several corpses, and outfitted his own customized garb, the mottled fallen patches of thick wisent fur in the summer, atop sewn clusters of dried and variable mosses.

27 Alice's Rabbit Hole

There were few equivalents to such natural trappings for keeping warm, even among the finest outdoor gear. Uncomfortable, much like a religious hair shirt, but moss and fur, hide and dried ferns could certainly stave off the onset of autumn freeze, and one might – theoretically – continue poaching into the winter, without needing fire, thus keeping underneath every radar screen including Google Earth, or the occasional satellite data, or very rare drone, utilized by park officials.

Chapter 28
Tuber aestivum

Forgetting the monk theory, Taman rationalized that perhaps this strange enigma in the forest was, after all, a lost member of a late fall mushroom-gathering party – nothing illegal in that, were it fifteen kilometers to the north and the east. But here he was. Deep into the off-limits zone of the National Park, which – on the Belarus side – constituted a very serious offense. Nobody but Taman and a few other scientists were occasionally allowed into this part of that wilderness. Indeed, no other scientist had visited in over three years. Trespassers would be fined heavily, jailed, or even shot, under any number of pretexts – including spying – should the military guards get involved.

Was the man truly discombobulated? He would have obviously spent several nights out in order to have reached so remote a patch by early dawn light.

Maybe Taman should have gone out to help him. The man held no weapon that Taman could recognize. No backpack. Not even the theoretical Bowie knife he might have used to fashion such disguise from other animals. If he had poached a wisent, well, he could expect a lifetime of servitude. But there was nothing to distinguish him as a bad man with a motive.

On the other hand, if this were the rabbit… No, he mused. Not a chance. The person evidenced no discipline or resolve, no digging tools.

Neither did Taman care to engage a stranger at his site of special scientific importance. His three graduate students were dependent on the data deriving from these coveted coordinates for their doctorates. Their work was akin to a military op. Not even other scientists knew the precise location of the monitoring plots. Let the man fumble and forage. He'd eventually find his way out of the labyrinth. This was not Alaska, after all. The poor bugger, scrounging for mushrooms to eat, was bound to be swarmed by ticks soon enough. That should discourage him from a lengthier sojourn come summer.

Within an hour of the intruder's dissipation together with the early morning fog, Taman investigated. He knew at once what was going on. The monitoring apparatus had not been touched. None of the control plots showed even a footprint. Nothing to indicate the presence of another. Twenty feet from the 80-foot experimental perimeter, the man had been digging for truffles.

A die-hard fungivore. Probably on his secret way right now to a kitchen, a saucepan, and an autumn brunch of burgundy truffles, *Tuber aestivum*, served up in chopped garlic with brandy. Or, preserved for auction. After all, two years before, the astonishing price of $330,000 had been paid for a mere 1.3 kilograms of rare Tuber.

These were the many conflicting thoughts that entered Taman's mind. Days later, following his solitary time-out, with many idle hours to ponder the mysterious intruder, scrutinizing every aspect of the images of the man captured on Taman's Nikon, it began to rain on him. Serious rain. Real evacuation. Heavy storms and intense lightning. As in any forest of old trees, the greatest danger was not the *de minimis* venom of a common European viper, nor even the tick-borne pathogens. Rather, a falling tree that might suddenly collapse without warning, especially during intense downpours and lightning, coupled with a rift of sudden snowfalls back to back. Freaky meteorology.

A foot of sleet in one hour. Impassable bogs becoming a single impenetrable flow of dark brown execrable waters, unless you knew the territory and had an instinct about which logs were floating and which were actually rooted to something sturdy. Ice was still plastered to the forest, making for a lethal combination. Micro-climes within Bialowieza were bizarre; temperatures varied from one shadow to another as if it were a different planet.

It snowed, then rained, then snowed again the entire time Taman drove his old, but sturdy Volvo the six hours back, nearly 400 kilometers, to his apartment in Minsk. Several nights at his computer, saying nothing about it to his wife, Sofia, who was quite occupied with her own students. She taught reading to children and had done so for over a decade, seeing her role in Belarus as preparing her charges for the new century.

Reading, she knew, was everything. Reading real books, one every night for the six-year-olds. Doctor Dolittle. Beatrix Potter. Bambi. Wild Animals I Have Known by Ernest Seton Thompson.

And their parents could not simply sign a form saying their child had done so. Sofia required more proof than that. Her kids were stars, as a result. And their two sons, Taman's and Sofia's – Alexander – a possibly quite talented back-room guitarist à la Bob Dylan, but also a kid with a curious interest in chemistry (he could not help but admire someone like Alexander Porfiryevich Borodin, the great Romantic composer, doctor, and chemist); the other – Misha – a prodigious wormhole buff with a supreme penchant for anything involving sines or cosines, a disciple of the late Nobel Prize-winning bongo drummer, Richard Feynman.

His math came to him without an iota of study; physics in a heartbeat; astrogeophysics by intuition. He was, in other words, the stuff of a Cal Tech or MIT full scholarship. That was his parents' dream, while Taman, absorbing one revelation after another in his out-of-date shambles of an office, was finally driven by astonishing turns, to the forbidding Rhombi. Into the bowels of a time in history that had transformed his own family with such intensity that, at the moment, he felt waves of paranoia stilling the air around him inside the heavy density of the National Library.

And then, piece by piece falling incredibly into place, he had gone with true panic to see his Great-aunt Sarah.

Chapter 29
Pavilion 3

7:40 p.m. The voice at once struck David Lev with the frightening ring of all that was familiar in another time, another world, that he had endeavored long ago to bury.

"Professor Lev?"

Pavilion 3 was still buzzing with ephemeral foot traffic, tired of continuing media interviews, and coverage of myriad stars in their own rights, or stars for five minutes of camera time, at any rate; most representing the BRICS – Brazil, Russia, India, China, and South Africa; and mini-BRICS – Mexico, Turkey, South Korea, the Philippines, and Indonesia.

"Who is this?" demanded Lev into his now 20% charged iPhone.

"Techsupport. I have been trying to find a way to get in touch with you for some time. Sorry about the hijinks."

That added a kindness, a nanosecond of intuitive comfort. Nonetheless: "Techsupport? What the hell is this? All those posted cryptic annoying stalker-like messages. What is your name? What do you want from me? Do I know you?"

"No. We have never met. But we must, and I say this – I hope my English is sufficient?"

"Your English is fine."

"Thank you. Because it is more important than you can possibly know. Trust me. Very important. Family. Your family. Our family. The War. World War II. You see? Yes. That far back. But now. Today. Right now, I am at a café in Leblon. It is owned by a relative. His friends, they are here for you, also. We're safe here."

"Safe? I don't understand. Who is we? And 'here for you'? What the hell does that mean?"

"Please. Trust me. Others are not so good. Drug companies. No more to say, not on the phone, please, just trust me."

"My computer is working fine. I don't need techsupport. And if you are a pimp, well, I'm sorry. I'm well into my eighties, and married and quite happy. Got it?"

"Dr. Lev. Please. No matter for joking. You are not listening to me. Listen to me. Do you really think I would go to this much effort? Of course I have thoroughly studied your early paper on Wallace. That's the lucky coincidence. My cousin works in the bar."

"Got it. Not exactly the same Alfred Russell, is it?"

"Quite right."

"World War II. I put all that long behind me. So what is this really about? Money? You've finally solved Darin and Wallace's great riddle"? Lev felt the unerring stirrings of a panic attack, the terrifying audible palpitations of his heart. He knew the signs only too well, from high school years, a pulse rate exceeding 140, blood pressure up to 170 over 115. The diastolic, of course, was the potential killer.

A panic that just burgeoned out of nowhere. Usually it was combined with some exploding, totally unexpected image of his childhood. But any kind of stress brought it on, in the guise of fear. Fear of death. Fear of pain. Fear of collapsing. Fear of being shot, stabbed, or a loved one disappearing in the middle of the night.

Even of financial ruin. He couldn't explain it. None of it was entirely logical. But it all came down to the one word that melded those millions of descendants of the most gruesome, forever unintelligible outrage of the modern world. Lev was, however distant and removed from its worst psychological epicenters, an inescapable *Survivor*. Forever.

"I cannot speak freely on a telephone. Please believe me when I tell you I must see you in person. I really am a great admirer of your scientific career."

"That's all well and good but I am on a plane headed back to the U.S. tomorrow and I do not have time for whatever it is. Where are you from? Your accent?"

"Professor, please. No questions. I apologize for that but I cannot overemphasize the urgency of our meeting."

"Enough bullshit. I don't keep secrets. Listen, whoever you are. I am heading to Leblon. My hotel is there. How's the food at your relative's café? I'll be bringing my Ph.D. assistant."

Chapter 30
Rio Maelstrom

By 10:15 p.m. Lev and Malcolm were standing on a sidewalk in the rain in front of a café almost named after one of his early heroes, but not quite.

He just wanted dinner, and to get this very uncomfortable premonition done and over with, liberated from his chest – it was the accent that had instantly hit him over the telephone with whoever the fuck "techsupport" was – as men and women poured in and out of the obviously popular nightspot along CEP Rua Desembargador Alfredo Russel.

He and Malcolm merged with five "kids" as they entered the queue for what, in Rio, amounted to an early dinner reservation. Lev was wet to the bone, and saw himself in a convex security mirror shivering "insensibly" like an Oliver Twist.

Nobody seemed the least interested in these two Americans being there. This was Rio at night, after all, during the biggest convention in its history. Lev caught the attention of the maître d', and he and Malcolm were seated at a table in the middle of the chaos and noise. Just perfect, Lev reckoned.

Lev heard three different pieces of music being played simultaneously – possibly two coming from adjacent cafés, but one was distinctly the famed Tarrega guitar piece, "Recuerdos de la Alhambra," not Brazilian, of course, but certainly at home in Lev's mind.

"What are you having?" he asked his young colleague.

"A glass of Merlot, Cabernet, I don't care."

Lev wasted little time scanning the menu. The two men motioned the waiter who efficiently took their orders. Two plates of *empadinhas de palmito*, okra instead of meat in a large *feijoada*. And a Guinness Extra Stout for Lev.

"Bring the beer first," he said. "And bring me a glass of ice."

"Certainly, sir," the garçom hastened gently in fluent English, not one to quibble with the requisite temperature for drinking a proper Guinness.

Lev looked at his iPhone, realized it was still early evening in LA and began to call his wife, when the garçom returned with a tray harboring the mug of beer, a glass of ice, and the 375 ml half-bottle of local Merlot. From behind him emerged a middle-aged man, probably around 51 or 52, Lev determined, and extremely haggard.

As the waiter repaired to his busier spheres, the stranger took the foreground. "I am techsupport. May I?"

Chapter 31
Euclid's Ghost

"And who am I?" Lev retorted, at ease with this nonsensical cloak-and-dagger, as he carefully poured part of his beer into the glass with ice, then helped himself unceremoniously to a healthy swig. He had only contempt for warm beer.

"Please, don't get up."

"This is my colleague, Mal."

"Malcolm."

"A pleasure, gentlemen. Thank you for coming. If I am correct, you, Dr. Lev, are from the Hajnowka-Narewka?" said the stranger.

Lev was desperately unhappy with this revelation out of nowhere. He put down his glass. "Whoever you are, Mr. Techsupport, I don't want to hear about ghosts, thank you very much. It's late and my friend and I would like to enjoy our dinner in peace."

"Peace? Dr. Lev, I do not bring peace, sorry to say. But maybe long-lost resolution."

"What?"

"Of course. To be expected. There is much to say, little time."

"Always like that."

"This is different. The stakes. I am at a loss to know how best to begin."

"At the beginning, if it's not going to take more than fifteen minutes. I'm tired. I'm sure my young friend is as well. We have a long flight tomorrow."

"My name is Taman. Taman Chernichevsky. My graduate students insist on calling me Doctor."

"What the hell is techsupport?"

"T.E.C. Taman Euclid Chernichevsky."

"Euclid?" Lev repeated, just a hair incredulous.

"My father admired mathematics. I know it's a bit weird, but beautifully so. You'd have to know him."

"What's his name?"

"My father is Isaac."

"Still alive?"

"Yes. Hanging on. Anyway, there it is, T-E-C. Tech."

"Uh-huh. So very clever. But what support do you imagine I need, and why all this stupid secrecy shit? Messages in the night? Are you trying to defect to the West? Hardly an issue these days. You are obviously not Brazilian. Russian?"

"No."

"What, then? Polish? I certainly detect it."

"Belarusian."

Chapter 32
The Moment That Would Change a Lifetime

Malcolm gave Professor Lev a curious look. He had run into quite a few Belarusians at the Summit and knew they represented the last bizarre bastion of isolationism in Europe, although there were numerous signs of progress, particularly in the realms of scientific collaboration.

"Yes – a half-dozen of us made it to the Summit," Taman replied.

"A totalitarian state, no?"

"That's the general rumor. But no, quite normal in most respects, just internally dependent on Russia. It is not the healthiest dependency, as you probably know. Have you been to Belarus?"

"Not lately," Lev said.

"Ever?"

"That's complicated. Maybe when I was four or five, playing in the woods."

"I understand."

"Really? How so?"

"I just do."

"Right. This conversation is already cloaked in a riddle, rendered feeble between, let's face it, strangers with absolutely no clarification. Why don't you make this simple for the three of us and tell me what the fuck you're talking about. What do you want?"

Lev finished his beer and ordered another. Malcolm observed the stranger carefully, taking his time with his merlot.

The stranger, Chernichevsky, was, by his every gesture and rate and distribution of distractions – Lev could not help but perceive behavior as an ethologist, with its own language of verbal and non-verbal attributes – clearly in a kind of semiotic purgatory, signaling more than he was saying. This duplicity did not help a meet-and-greet late-night encounter.

"Totalitarianism is not entirely misfitted as a term to my country. As a scientist I must confess, we have issues, and a few in particular now bearing down on what has made my seeking you out so critical."

"No clapping allowed in Belarusian public, is that true?"

"Uh-huh. Of course, it's illegal for children to climb trees in Singapore, so there you go."

"Yeah. That sucks."

"A lot of famous Belarusians in the States who couldn't wait to get the hell out of there," Malcolm chimed in.

"We are well accounted for abroad, but name a country that has not seen its flow of exiles."

Lev took another swig. "That is true. Want a beer?"

"Yes, I think so."

The garçom was summoned, and Taman was appropriately supplied. Malcolm poured himself the remains of his half-bottle of wine.

"So you've been at the Summit? Did you give a presentation?" Lev inquired.

"No. But I did hear yours, over the monitors. Excellent. You referenced Eastern Europe."

"Yes. A region that is certainly part of the general demise of the Earth, wouldn't you say? Antarctic glaciers melting at a point of no return. The Anthropocene, my friend. Most exotic birds, large mammals, habitat in general going extinct. I think you understand."

"I couldn't say."

"Come on, sure you could. Any kid on the block can sense it coming."

Taman noticed how Lev's young assistant rather rolled his eyes.

"You were certainly not entirely upbeat, unlike many of the presenters, that's for sure," Taman continued.

"No," Lev answered. "I've never been accused of upbeat. So. Belarus. That explains how you would, at the very least, be familiar with that region which for some time has been Poland. But it does not come close to clarifying how you know that I came from there over seventy-five years ago. And why the fuck would you care?"

Taman lowered his face. Malcolm easily noticed the tension rising to the surface. The table itself seemed to be hovering beneath his eyes and this stranger from Belarus appeared to be absorbing all the fear in the world at the moment.

The beer had already gone to the Professor's head. It had the effect on him of speaking his mind with fluent disregard for the least civility... "fuck" and "fucking" and "what the fuck..." Quite the linguist.

Continuing: "And why here, now, like this?"

Taman had rehearsed this over and over. But now, in the face of its reality, in the middle table, the noisiest spot in the restaurant, as planned, his half-whispered speech came to near naught – nor had he expected a companion with Lev, although it was clear that Malcolm fit the perfect stereotypical grad student. Taman would know.

"Dr. Lev, I am a mycologist."

"That's nice."

He waited. The middle-aged man, Taman, was fidgeting awkwardly. With his tousled hair and nervous sweat bearing down, he'd been discretely scanning the restaurant continually, to all sides, noting that a certain individual – it was his cousin – was watching after him, as reflected in the mirror above the long bar with its dazzling collection of colorful labels, every liquor in South America and more. Campo de Encanto Pisco, Ypióca, any mixologist's paradise of capirinhas.

"So what is this?" Lev asked. "What am I doing here talking to you? I've got to pack. Call my wife. Check three hundred emails. You are not part of the equation."

"I think I might be, as of this moment."

"Really?" Lev speculated, contemplating the nature of what appeared to be a bizarre shakedown. He made no bones about his annoyance, but certainly felt the threatening air. At his age, Lev suffered nobody, said precisely what he thought. In radio interviews he would never hesitate to name names or break silence on universal truths with a savage clarity.

Once, on the Huffington Post live webcast, he'd uttered a downstreamed "Who's Who" of the best lawsuits waiting to happen. But, not unlike a stand-up comedian, it would be hard to take a David Lev to court: his science was impeccable, his age offering up the immunity of Socrates; the cantankerous nature of his verbal outrages so obviously addled as to verge on a kind of public persona senility, or, PPS.

"Sorry, what was your name again?" Malcolm asked, finishing his meal.

"Taman Chernichevsky. And what was your surname? I think I missed it."

"Howler."

"As in the black monkey?" Taman asked.

"Apparently. And Red."

"And Brown," Lev added.

"You are certainly in the right country," Taman volunteered with a vague grin.

"Yeah, I'm told there's about nineteen subspecies throughout South America."

"First time in Brazil?"

"Yes."

The small talk defused the fact that Lev now saw precisely the figure of the man Taman had been keeping an eye on, and vice versa. Two men, in fact.

Lev was still struggling with the near threat uttered seconds before by the Belarusian enigma.

...I think I might be, as of this moment...more than fucking spooked. The whole tone of this, laden with gaps, disparately injected, unfriendly warning signals, in retrospect, would all seem like some kind of nightmarish Déjà vécu, Lev's thoughts eerily spinning backwards, now.

And it was.

Chapter 33
Ugly World

…Crashing and screaming…everywhere in this world of bright bottle features detonated in a meteorite shower as Lev felt himself flying, thrown upwards from his chair, suction blowing outwards, an awesome explosion wracking through the café, glass and metal exploding like an avalanche.

People were hurtled to the floor, among them Mal; an emblazoning of sudden desperate smoke, and the burning blinding dullness in the mindless instant of disaster.

"Mal?" Lev muttered with bare breath.

No response.

Chapter 34
Daze

11:59 p.m., No bairro do Leblon. "Where am I?" David Lev mumbled. He could feel things on his body; on his face, his arms, his legs. Even his back. Pain. The etymology, as it were, of consciousness.

All around him, there were sounds. Beeping noises. Sirens. Shuffling of staff. Disorientation.

"Hospital Municipal Miguel Couto," a nurse replied, checking the IVs that dangled and rose, both, from his right arm and his beer-drinking hand.

"That stings."

"I'm sorry." She took pressure off his right arm, shifting the fluid line. "You are lucky, I think. Some were not."

"Leblon, did you say? What… what happened?"

"Shhh…"

"I'm telling you I'm all right."

He tried to sit up. The nurse rested her hand gently on his right wrist.

"A doctor and anesthesiologist are on their way."

"Where are my things? Where's Mal? What happened?"

Lev had once lost his passport in Haiti, a 90-minute flight from Fort Lauderdale. It took him nine days to get back to America.

He heard only the mussy sounding "Com administradores e polícia…"

"But what, where…" and he stopped himself, peering back at that moment. A part of the compounding pharmaceutical computation… That was Lev thinking. He tried to turn his head to see who was coming when he felt the surge of pain shoot through his left shoulder. He let out a yelp.

"Please," the nurse said with not a little distraction, as other victims were being rushed in.

"Você vai ficar bem."

"My Portuguese sucks. What's with my shoulder?"

Within minutes the pre-anesthetic was working its effects and Lev hardly realized that two men had just entered the room in which he lay weak, ineffectual, and without the slightest inner plan; men wearing blue gloves, masks, white frocks.

He felt a second injection. As he faintly pondered vague verbal instructions. He started to follow the game plan... 100, 99, 98... and then the two men and two nurses standing by, removing an ugly, splayed bullet, like shrapnel from the left shoulder of the inert and unconscious professor.

Chapter 35
Obfuscations

At the Municipal Hospital, a man dressed in a Gopnik-like jogging suit, hooded sweatshirt, running shoes, resembling an Os Tupis rugby-type guy or footballer, shuffled past the growing surge of visitors, doctors and investigators. He spoke broken Portuguese to a nurse enquiring after someone from Belarus.

"Você é um parente?"

"Sim."

She quickly glanced down at her paperwork, then shook her head.

"Sinto muito. Eu não posso te dizer nada."

"Que tal um cientista americano?"

"Sinto muito." And she rushed away towards more pressing duties.

The footballer scouted the harried scene. He heard some English amid numerous languages and gathered as close to its source as possible, taking a seat among a motley row of what were most likely relatives waiting for the little news that was, as yet, being parsed out by hospital authorities.

He saw a tall gentleman, an American, tie, no jacket, white shirt sleeves rolled up, in jeans, and a badge marking him presumably as Embassy, speaking to a policeman positioned quite purposefully in the hallway.

The policeman was on edge, not happy about speaking in the open, then noticed the fellow in his jogging clothes sitting down the hall among others. He fit in, but the policeman also saw how the guy looked down in sync with the meeting of two sets of eyes. He might have ordinarily taken a closer look, but events were occurring too rapidly, the officer's cell phone vibrating, and the American hurling one official question after another, in English.

The men parted. This was all absorbed by the footballer, who tried to follow the American as he entered an elevator heading up several floors.

The man moved quickly to the stairs, sprinting up past opposing traffic, two stairs at a time, moving onto the second, third, and then fourth floor. He was just able to see the American heading into a particular room, flashing his credentials at the two policemen who stood guard there.

"Professor Lev?" the tall, strapping American said, entering David Lev's private room. A doctor was there conferring with Lev, and the beige curtain was half open.

"And you are?"

"From the American Consulate. John Vespers." He showed his picture ID. "Commercial attaché."

"I'm told they've removed a bullet from my shoulder. It missed the bone, happily enough," Lev declared.

"He was very lucky," the doctor added. "But it is a police matter."

"Actually, as your patient is an American citizen…"

"Yes, yes, that's fine. Are you going to be all right for the time being?" the doctor asked Lev.

"You guys have been great."

"Não se preocupe. É o nosso trabalho."

"You understood?" asked the exceedingly calm Vespers.

"Well enough. I just am not quite sure about this news of a bullet. What happened?"

"That's what we all want to find out. May I?"

"You're here. Where's my younger colleague?"

"Who would that be?"

"Malcolm Howler. Mal, for short. My Ph.D. student. Is he all right?"

"I really don't know. I'll go ask." He went out the door.

…fuck…Lev contemplated… Commercial attaché… What is that all about?

Lev waited, searching for his iPhone. Gone. For his wallet, also gone. Indeed, he had no possessions whatsoever. Only the backwards hospital gown in which he was all tied up, among IV tubes and pieces of tape on multiple body parts.

Vespers re-entered Lev's room. "I'm afraid I can't divulge his condition."

"What do you mean you can't? You can't or you won't?"

"Brazilian laws."

"Bullshit. We're Americans, and he's my friend."

"Professor, they're operating on him as we speak."

"What happened?"

"We intend to find out."

Vespers sat on the chair the doctor had occupied, then closed the curtain.

"I'm not the consulate-general, but the duty officer assigned to the incident, for now."

"Commercial attaché. That's what your badge says."

"Yes." Vespers made nothing of it. "We all have to live with our little titles, however fleeting they may be."

"Well, I'll tell you right off, no one ever profited from any of my discoveries."

"How unfortunate," said Vespers. "Nothing wrong with good old America commerce."

Lev said nothing. Too inane, he thought, to reply. He already felt in his gut an immense distrust of this guy, whoever he really was. Lev had been eyeing a clock that was on the wall and contemplating his situation. He replied, "And I'm flying home in about seven hours."

"You live in Los Angeles, I gather."

"Not a question. You know who I am, then."

"That's my job. Listen, Professor,"

"You've got my full attention."

"Professor, do you know what happened in that café?"

"Chaos. Something went bang in the night. And I feel like shit. I need to pee which means I've got to get rid of this IV. And I need my iPhone. And you need to call Malcolm's parents. I think they live in Montreal. And his fiancée, Militia. I mean Melissa. Somewhere in LA. She's with the LAPD, I believe."

"Really? LAPD?" Vespers withdrew a pen from his shirt pocket, a notepad from his satchel, then wrote down the information.

"You seem surprised?"

"Terrorist attacks are always an unwelcome surprise, Professor."

"Yes. I really gotta pee."

Lev tried to find the right button to call for the nurse, but could not.

"I got it," Vespers said. He went out the door.

The nurse immediately came into Lev's cramped room pushing a wheelchair prepared to take the addled Professor to the bathroom, when another nurse came forward, speaking quickly with a policeman who stood adjacent and helped carry Lev's things in a bag. Lev had not noticed them before. They'd been under his bed. The policeman nodded to the two American cops guarding the room, came in and left the things on the chair beside Lev's bed where Vespers had been sitting.

Lev tried to notice everything, but it all went down too fast. The whirl of existence enshrouded him. His head was burning up.

It was only then that the professor actually noticed that his hospital room was indeed being guarded by two American policemen.

"Good morning," he said to them.

"You too," one of the cops mumbled.

In the bathroom Lev felt himself on the verge of a panic attack. Blood, he noticed, dripping on the floor from the PJs of his left leg. What the fuck is this? His atenolol and ativan pills were in his bag, on that chair beside his bed.

Outside, the nurse waited, then spoke through the door, "Dr. Lev? Are you all right in there?"

Anyone within thirty feet would have heard that.

Lev managed to correct his free fall, stood up, stared into the mirror: "Fuck!" he said aloud. Then shouted it a second time.

"I need a paper bag. Get me my things, on the chair beside my bed, hurry!"

Back out in the hallway, into the brutish steel chair, back to his room past the distinctive mess of humanity all around. Dozens of other patients, maybe other victims, he couldn't tell, didn't really care anymore.

The nurse entered and Lev rapidly rummaged through his things, popping some pills. The nurse turned away, not wanting to embarrass him.

Back out in the hallway, the nurse helped seat the professor into the wheelchair. Vespers was on hand.

"Give me a minute, I need to call my wife," Lev said, his throat hoarse. He was choking on one of the many panic attack pills he'd just downed with water from the sink in the bathroom. He normally maintained a rule against drinking tap water abroad. This was the exception.

"You all right?"

"I get panic attacks," Lev declared, leaning his head on his hand to one side of the wheel chair as the nurse wheeled him back to his hospital bed.

"You realize it's after midnight in Los Angeles."

Lev shook his head, as the phone was ringing. He'd forgotten that he just tried to call her.

His wife, Sasha, near instantly picked up.

"Sweetheart."

"I saw it on CNN," she immediately volunteered. "That's the area you're staying in, isn't it? Leblon?"

"More or less. Look, I'm fine. I'll call you when I'm in a taxi from LAX. You all right?"

He wrapped it up. He did not want to go into it. "I love you."

He noticed dozens of unanswered emails on the iPhone but first checked through his things. His wallet and passport were secure, along with other documents inside his briefcase.

Once back in his bed, lying there out of breath, Lev looked up at Vespers, who was writing something down.

"What are you writing?"

"We're just trying to get a handle on all this. What do you remember?"

"I told you. One hell of an explosion. Just when I was enjoying a really good cold beer. Threw me off my chair. Heard a lot of screaming."

"And you were with two friends?"

Lev was furious and made it clear: "I told you! Malcolm Howler, look him up. A grad student at UCLA, working on his Ph.D. for three years. I brought him here on an NSF grant to be my assistant, write that down. He's Canadian. I told you."

"The other guy. Who was he?"

"He was a stranger to me. Totally. What do you know about him?"

"Nothing yet. Video surveillance. We've been through the footage."

"We?"

"The consulate-general and a few staff. You'll meet her. Yes, there was an explosion, planted in the area of the bar. Investigators are pretty sure it was a diversion."

"From what?"

"Do you have any enemies here in Rio de Janeiro, Dr. Lev?"

"As in anyone who would like to see me gone? Yeah, me for one. From this fucking bed."

"Any enemies, professor?"

"Plenty. But governments have been ignoring me for decades, or trying to. Nothing new about that. So the answer to your riddle is, no. I was one of the first tree-huggers, you probably know that though. And remember Rocky Flats? No. You're too young."

"Dr. Lev, only three bullets were fired, from what we know thus far. All in your direction, you and the two gentlemen with you. You were very lucky. Whoever did this got their timing slightly awry. And their aim."

"At me?" Now the psychosomatic reverberations angled downward from the shoulder that had evidently been struck, a single bullet missing the bone narrowly.

"Yes. Give me a minute."

Vespers went out to the nurses' station and returned within a few minutes carrying a small tray on which had been placed a forensics plastic bag.

"This is the bullet they removed from your shoulder two hours ago." Vespers showed the professor the bullet.

"I can't tell you the caliber, or let you touch it. It's evidence."

"NATO 5.56 millimeters would be my guess, from limited experience," Lev said.

"I really couldn't comment. I just thought you'd like to see it."

"Bigger than that from a BB gun," Lev ruminated aloud, thinking back to a juvenile crow with a BB he once rescued, probably shot by some kid, that had shattered the beautiful bird's tibiotarsus, rupturing muscles and flesh from the peronaeus brevis to the patellar tendon.

"Yes. The initial indications, movement of the blast, versus the direction of the bullet. It's forensics. Quite obvious choreography. You and the stranger were the targets. So who was he?"

"That's a very good question. Is he all right?"

No response.

"I presume you've spoken with him?"

"We can't find him, not yet. He was released from this hospital two hours ago. They screwed up. But he apparently walked out of here on his own. Bomb blasts have their own fickle way of hitting some targets, missing others. When you were in the bathroom I learned your student is going to pull through. Look, in about twenty minutes, just before the hospital does its breakfast rounds, an armored embassy car is going to arrive to take us together to the consulate. I doubt that you're in any condition to fly home today."

"Wrong. My wife needs to visit her sister up in the Bay Area. She's got the big C. We have two parrots I need to look after."

"Your parents? Long-lived. For a minute I thought you said parrots."

"That is correct. Parrots."

"Birds?"

"Yeah. Parrots are birds."

"Surely there's a boarding facility in Los Angeles?"

"No."

"Dr. Lev, two people were killed in the blast, dozens injured. They were gunning for you, and, we believe, whoever was dining with you, and I don't think your graduate student was the object of their wrath."

"We weren't dining. If you saw the video you know that. Mal and I were drinking a beer with someone I had never met before. A scientist from the Summit. Mal called him a stalker, as a matter of fact. Which rather spooked me, I must say. How many times do I have to repeat myself?"

"You've not repeated anything. We're just talking here."

"I'm just a little perplexed by all this."

"Naturally. So, from which country?"

"I'm an American, obviously," Lev said with sharp impatience.

"I meant your friend?"

"Not my friend, goddammit. I don't know who the fuck he is. He said Belarus."

Vespers wrote it down. Then, glancing up, "What name did he give to you?"

"Oh shit... It'll come to me. Maybe there's a touch of amnesia. I'm not accustomed to meeting strangers just before explosions." He tried but could not put the name together in his head. "If he was here, then surely his name was obtained on admission?"

Vespers again shook his head. Another screw-up in hospital protocols. "Now listen to me. The local police will have a few questions. They agree that it would be better to chat with you at the U.S. Consulate."

"But this is crazy. I gave a speech yesterday morning. Nobody threw any rotten tomatoes. Well, a plastic tomahawk."

"What?"

"Just a pissed-off local, an Indian from the Amazon expressing utterly justifiable contempt for the whole proceeding. I mean, let's face it, since Columbus, and then Pedro here in Brazil, they've all been screwed. Every indigenous person. The largest demographic demise in human history. One hundred ten million native peoples were massacred."

"A hundred and ten? What are you talking about?"

"Do your homework. That's right. Add that to World War II and well, anyway, like I said. What was I saying?"

Vespers stared blankly at the professor.

"Oh, the tomahawk. It was a toy. Personally, I would have used bronze. But there was nobody to throw it at, in the end. Nobody and everybody. Anyway, fuck it. Made his point, damn right. What did you, what – what were you asking me? I lost my train of thought."

"If you might have any known enemies here in Rio? Perhaps your research? Were you collaborating with the man from Belarus?"

Lev did not like the line of questioning, not at all. Alarm bells ringing everywhere. The whole goddamned picture stank.

He then thought how best to proceed, what to say, what not to say. Everything had been made difficult.

"You a chain smoker?"

"What?"

"I never cared for Tchaikovsky's 'Suite for Orchestra No. 4, the Preghiere'."

"You've been heavily drugged," Vespers said with a tone of conciliatory bullshit.

"That's it. The drugs. I was anesthetized, then?"

"I'm told they performed minor bone surgery, yes."

"That accounts for it."

"Right," Vespers said, realizing what was going on.

"Considering it was just your normal apocalyptic diatribe, dismal data scattered here and there – my speech at the Summit – I'd go so far as to characterize its reception as verging on the rapturous. Of course, there was a hell of a lot of security everywhere. Excepting the plastic tomahawk."

"They don't usually have terrorist attacks in Rio."

"No. They reserve them for the rainforest."

"Who might have known you'd be at that café?"

"Only the fellow from Belarus. Possibly the bartender. Oh, a second person. I'm pretty sure."

"Male? Female?"

"Male. And there were these weird little messages for me at Rio Centro."

Vespers withdrew a little sealed plastic bag from his shirt pocket. "You mean one of these?" There was the wadded Post-it. "TechSupport? Was it that?"

"Hell if I know."

"He knew you were staying in the area. How?"

"I have no idea."

"Only you can make this easy. It'll expedite your getting out of here."

"Cooperate? Do I look like I'm not cooperating? I didn't go through your stuff. And you didn't get blown up and shot. So go fuck yourself."

"Sorry, professor."

"Look, I was once controversial among maybe fifty colleagues worldwide. And even that is self-flattery. Now, they're mostly retired and arthritic, or dead. The worst I've gotten are a few nasty blogs and the proverbial one-star book review. It doesn't exactly add up, does it? The guy's name will come to me."

"We know what he looks like. The embassies have gotten the image."

Lev's mind suddenly raced to the words... part of the equation.

"What was the book about?"

"Book. What book?"

"You just said something about a book of yours?"

"I've written dozens of books. Pretty much all about the same thing."

"And what would that be?"

"Human nature. The end of the world."

"Well, there you go."

"Jesus. Are all you embassy people this perverse?"

"It's five in the morning. I'll give you that."

Chapter 36
Early Morning, Avenida President Wilson

The SUV came roaring from across the intersection straight at the vehicle carrying Lev, Vespers, and two policemen just as they turned onto Avenida Presidente Wilson.

It slammed into the side bearing Vespers, and the telling fire accompanying the collision would have proved disastrous had the Embassy's vehicle been built of anything less sturdy.

Two well-armed men were sprinting fast in different directions. That much Lev could make out through the tinted glass,.

Another man, from behind a vehicle parked across the still-quiet downtown area was shooting at them, and then dropped his weapon and ran in yet a third direction. He was the footballer, his hood now down, long black hair flying in the June mist.

Chaos prevailed.

"GO! GO! GO!" Vespers shouted, as the driver backed up, slamming one of the assailants who was on foot. He was firing rounds from an assault rifle directly into the rear windshield without penetrating the glass as the Embassy SUV ran directly over his body. Lev heard the anatomical crunching of a mammal.

The SUV careened alongside two parked vehicles and continued at high speed towards the Consulate.

"FASTER!" Vespers was working two iPhones.

There was a special emergency way in, laser gate, sirens and red lights flashing, Marine security guards targeting all incoming, machine guns aimed and ready.

Lev, his shoulder having been swung out of whatever painkillers had earlier kicked in, now watched the slow-motion ordeal with a replay button in his weary eyes as he was shuffled, surrounded and moved in a military maneuver that got him inside the Consulate corridor, then into a secured elevator going up.

"Hit it again!" Vespers ordered, as one of the military escorts pushed the floor number again and again. The elevator was not taking orders.

At that moment the escort, fully cloaked in U.S. Marine attire, turned and raised a weapon towards the professor. Without a microsecond's delay, a second marine nailed him through the cranium with his own easily deployed handgun. The impostor had been sloppy, slowed by the elevator glitch; his senses delayed that hair-splitting instant that enabled him to be killed, his head splattered all over the elevator's interior, and all over the professor's, Vesper's, and the apparently real marine's face.

Embassy officials were flying past them as they emerged into the inner sanctum where the professor, Vespers, and the marine spilled out, wiping the blood and tissue off their own faces to the extent possible.

The elevator was secured by waiting military who knelt beside the dead man, locking the door to keep it open and the elevator from going back down.

"NOW!" yelled yet another marine stationed in the hallway.

Others came running to the assistance of the bloodied threesome who were led rapidly down an adjoining hallway.

They reached an open door, poured through, and that door was locked behind them, three men guarding it.

"Jesus," Vespers groaned, almost too consciously.

"Hmm. The U.S. Embassy. What a relief," Professor Lev said with unusual sang froid.

"This is Ms. Chambers, our Consul-General," Vespers exclaimed.

"Professor," she said by way of a greeting. "I'm so sorry for all this."

She started to shake his hand, but there was blood dripping from his face.

She looked at one of the marines, still locked and loaded. "Please get Dr. Lev a clean towel. You too, John."

They both had remnants of guts and blood on their faces and necks.

Her speakerphone was on when a voice almost instantly addressed her, as she was taking her seat. "They think they have a match on the body."

She turned the speaker off, and raised the portable phone to her ear, then kept speaking. "Uh-huh. Yes. Thank you."

Then the Consul-General, putting down her phone, removing a pen and writing just a few words, looked up at Vespers, pointing at her own forehead.

Vespers took a while to understood her sign language, and then removed a handkerchief and wiped off a bit more stuff from his own left temple. A human head, when it explodes at close range in an elevator, leaves one hell of a trace.

Chambers glanced up at the marines and local policemen who were assembled in her office, and then at Lev, before fixing her incredulous gaze upon Vespers: "What just happened?"

Chapter 37
The Bartender's Apartment

Taman Chernichevsky sat watching the news with his cousin, the bartender, at an apartment just three blocks from the café that had now been cordoned off and had been descended upon by half of Rio's security forces and police investigators. There was a knock at the door. Alberto took his handgun from under a pillow on the sofa where Taman sat nervously chewing on beef jerky.

Alberto walked over to the door and peeked through the hole. It was the footballer, out of breath. Alberto let him in.

"Your American is at the U.S. Consulate. His luck has been with him today."

"I presume you have a plan?" Alberto said, putting the gun back and casting a worried stare at his first cousin that betrayed his lack of confidence.

"I've got to reach him somehow."

"Let me see your phone," the footballer said.

"Why?"

"Just give it to me."

Taman handed Vincent his Android.

Vincent looked at it, got up and walked into the kitchen, opened drawers, took out a chopping board, found a cleaver, and then proceeded to destroy it.

"What are you doing?" Taman hollered.

"They knew exactly when and where he'd be," Vincent declared in a voice of absolute certainty.

"There was a video capture system above the bar. But I wouldn't bet on it having been destroyed in the blast," Alberto chimed in.

"You need to get out," Vincent said. "Jobim's less than an hour from here, if we leave this minute." He looked at Alberto, who nodded.

"Better to leave my car in long-term parking." Alberto tossed him a set of keys.

***

At Customs, Taman waited calmly in line. When it was his turn, he reacted with that quiet, dignified calm meant to counter the reality of his plight. The queue was comprised of his peer group from the Summit, from everywhere, which served his purpose like no other circumstances could have. The official looked him over, checked the monitor with exceeding patience, then with a distinct air of tired disinterest stamped the passport.

Taman proceeded to his gate, having purchased a ticket with a credit card over Alberto's computer. There was no other way, and only one seat left, even in First Class. Alberto had pulled his weight, to be sure, because it would have taken Taman more than a year, at his salary, to pay off such a fare. Alberto would make it on tips alone in one month.

Had Taman arrived even ten minutes later, had the line been longer, or had there been the slightest delay in early morning traffic – a dozen different potential scenarios – his picture would have popped up on the Customs Watch List and he would not have been sitting on a Lufthansa flight bound for Munich, leaving Rio in the rain of an early morning on June 23rd.

Taman asked only for ice water and a hot washrag from the attendant who was giving out champagne at 8:00 in the morning to those in his seating group. The flight lifted off, and Taman closed his eyes. He needed no convincing from Vincent that one of Taman's graduate students – and Taman by now was certain which one of them it was – had tipped off some very bad people.

Chapter 38
From the Knee Down

It had not just been a bad dream. The heavy air had not dissipated. And the mystery had never been officially explained.

There was a report of a terrorist bombing at a café in Rio, but no terrorist group ever claimed responsibility. No culprits were ever apprehended.

The horrible explosion was chalked up as one more unsolved mystery at the time of a run-up to the elections, rising unemployment, declining economic stability, and a turbulent political campaign in Brazil. Not to mention the international focus on the UN meetings. A perfect opportunity for the best and/or the worst of humanity.

***

Nearly three years would pass, giving Lev plenty of time to reflect on the precarious and personally dire meaning of the UN Summit. For all of the obvious time and expense, more than 50,000 participants, hopes, and dreams, very little had been accomplished. It was, he remarked somewhere in the press, a mere "footnote in the annals of the planet."

The Paris COP21 climate talks would take place in less than nine months, and the UN/IPCC Lima Summit had prepared groundwork for Paris that was likely to be undermined should the Republicans take the U.S. presidency in November, 2016. Nothing was clear. The global Total Human Fertility Rate, 82 million and counting per year. The continuing acceleration of the extinction of species. Loss of prime habitat. Death of the oceans. Weather "anomalies," like 80 inches of snow in Buffalo in a matter of hours.

Still, following upon the IUCN/World Parks Congress in Sydney – and its 53 sessions – it became clear that the adherence to the so-called Aichi Biodiversity Targets #11, referring to the "Technical Rationale extended" from the Japanese COP/10/INF/12/Rev.1, in which "at least 17% of terrestrial and inland water and 10% of coastal and marine areas" would be protected, was falling short of the new needs of this generation.

Claims by the IUCN (International Union for the Conservation of Nature and Natural Resources, based in Gland, Switzerland) that there were, as of November, 2014, a total of 209,000 protected areas, covering approximately "15.4 %" of the Earth's terrestrial and fresh water regions, as well as "3.4 %" of all oceans, were ambiguous numbers, at the very least, especially in light of the virtual absence of what was designated "No Kill" areas.

Scientists were, meanwhile, advocating for the killing of some species to favor others – keep the salmon, but kill several thousand cormorants, for example. Exterminate the Puerto Rican Coqui frog that, along with the rhinoceros banana beetle was overtaking Hawaii, but federally delist the endangered pinto abalone (a snail) – between British Columbia and Southern California – against all the evidence. Both species beautiful to behold, a quarter-inch in size. Both representative – like beluga whales and Ohlone tiger beetles – of the chaos inflicted by a planet of pathetic jesters, all deity-obsessed decisions by people incapable of getting over themselves.

Lev had seen it all, and there were few governments that he hadn't pissed off; data were coming and going. Realities supplanted by new surrealities. With the price of oil plummeting to below fifty dollars a barrel, cannily manipulated by the Saudis, the dominant GOP lawmakers inside the Beltway were gutting EPA initiatives, and even Obama was backing away from previously vowed government incentives for alternative energy sources. Smog levels were intensifying worldwide as consumers, airlines, truckers, and everyone involved in hydrocarbons, were feeling the spirit of a circus.

Meanwhile, the demise of canaries in every scientifically studied mineshaft was a virtual given. Whole populations of giraffes, tigers, rhinos, elephants, and – most emphatically – the Actinopterygii ray-finned fish, were quickly vanishing from the planet. In the case of the latter, for one of three taxonomic groups encompassing the vast majority of all fish in the oceans, most notably the 23,000 known species of the infraclass Teleostei, extinction was looming large.

As for Malcolm Howler, he had lost his left leg from the knee down in the explosion.

At his wedding to Melissa, the retired Professor Lev and his wife Sasha sat in the front row of the Church of the Good Shepherd.

Dr. Lev had been interrogated far beyond the questioning he had endured sixty years earlier, he stated, when defending his own dissertation. But nothing ever came of his insightful suggestions, particularly the fact that there was someone not just in the Consular elevator, but involved in other ways. He had his own suspicions, to be sure.

Moreover, he was certain someone had saved their lives on the drive to the Consulate. John Vespers denied it, attributing the Professor's hunch to his heavily drugged state. After all, two hours before he'd had surgery to remove a bullet from his left acromioclavicular joint. The bursa sac had ruptured. The damage to the professor's shoulder was serious at his age, although it would heal to some degree.

But impossibly, irrefutably, nothing had come to pass of the gruesome event at that café, in legal terms. Lev could only regret having ever gone, let alone having invited his dear younger friend to Rio. At least he had had the pleasure, six months later, of handing the remarkably robust Mal his newly consecrated, signed, sealed, and delivered doctoral dissertation.

And that was that.

Chapter 39
The Ides, 2015

In Minsk, a city dependent upon Russian oil and gas, funding for research like that conducted by the now 55-year-old Taman Chernichevsky was in a state of total collapse. Or that was the rationale by which his superior at work had conveyed the bad news – a stubble of a man Taman had always rather detested; a little wimp who had made his way up the ranks at the National Academy of Sciences by means unknown to any serious researcher.

But in the end, and after a series of negotiations – for Taman had been at BNAS for nearly three decades – he was given six months to wrap it up. By the fall, his lab in the forest was being shut down.

Meanwhile, Sarah ("Auntie") had died at her Alzheimer's clinic. Stashed in a drawer amid other people's socks, and under a phonebook, she left a gift that would be claimed by its rightful owner in the nick of time.

Within two weeks of Sarah's death, Taman's father, Isaac, also passed away, peacefully at home.

David's sister-in-law had perished after a long battle with ovarian cancer. In fact, she had done so at a clinic in Switzerland, on her own terms. Sasha had gone to be with her at the end, bringing the specified flowers in full bloom, and a CD player. A deliberated death with dignity.

Lev had ensured that Malcolm got an immediate professorship in Lev's own lab. As for Lev himself, he was now officially retired. Retired from the university, retired from the world. He hardly even had the appetite to have a meal with his few close friends, and Sasha had good reason to be concerned.

And then it got worse.

On the morning of March 15th, the professor's updated iPhone received a text message, the first of its kind, from "techsupport," a phrase, an identity, a series of nonsensical and traumatic events he had tried day and night to put behind him.

"There's more to it than that." This was the sole extent of the text.

Lev was never a patient man, nor was he equipped to invest his frayed emotions in ghosts or apparitions. This text message was downright appalling. He thought, somehow, it had all ended that night, despite the terrible memories.

Lev called Mal's wife.

"I have no idea," Melissa Howler told him over the phone. She had taken Malcolm's last name in marriage. "As you know, I left the force after Mal's injury. He needed me at home."

"Is there anything I should be doing?" Lev asked her, trying by whatever logical means to bring this seemingly continuing nightmare to some kind of closure.

"I'll call a few friends and see if anybody has a suggestion," she said. "But I wouldn't hold out much hope. It's old news, by the standards of nearly every morning headline."

She was right, and the professor felt very old. His earlier scientific bravado – the stuff of an entire lifetime – all his anger at the way humanity was continuing to fumble, had no more traction than a grieving ant unseen on the surface of a vast desert.

Half a world away, Taman's three graduate students had gone on with their careers, one within BNAS, one back in Warsaw with his government – Taman was unsure of what the man's Ph.D. would buy him there – and then there was Petrovsky, the Russian, who never said much, despite having benefitted from the quality of Taman's generous mentorship.

Now, nearly three years since the events in Rio, that person had obtained his Ph.D. under Taman, and managed to hold down a part-time position – Taman heard through a mutual connection – within the same forest, leading the rare tourists on a bicycle path about thirty-five kilometers from the research station.

If one continued on that path, during daylight hours it was possible to cross over, with proper documents, into Poland. Taman saw him occasionally. He got his first sense of strange vibes when Ivan alerted him to the fact that he didn't trust Ulyana's boy friend. Nothing more was ever said about it.

Then events began to disintegrate.

© M.C. Tobias

Chapter 40
The Communiqué

Hence, the timing of the first communication since that fateful night three years before, at his cousin Alberto's café.

The text message provided no insights from the few colleagues that Melissa spoke to, with no particular guidelines as to what might be pursued. It was, as they say, a "cold corpse."

Lev did not mention this new development to Sasha, who had her own medical issues, and had succumbed to a nearly obsessive passion to clean out every drawer and closet in their home. She seemed to be subconsciously putting her estate – *their* estate – in order.

Then, on the same day as the text message, Lev received a second terse directive telling him to go to a computer at the UCLA Library research section, at 5 p.m. that day, then utilize what was known as a Shibboleth users' sign-in for the Web of Science log-in.

Lev knew exactly what to do. He called Mal, who was in the middle of teaching a class, but within minutes managed to answer.

At 5 p.m. sharp they were seated in a quiet corner, out of sight, as Mal walked Lev through the Waiver, Embargo, and Addendum portions of the e-share work-in-progress levels of the University system, which is what the communiqué required.

"So it's him, isn't it?" Mal asked the professor.

"Yes. And you need to go. No point getting you in over your head."

"Really. You think?" Mal said as he pulled a Phillips screwdriver out of his pocket, held it in his teeth while he lifted his pants leg, and set about to tighten a nearly invisible screw on his titanium artificial leg.

"How's Melissa?"

"Fine thanks, David. She sends her love. You sure you can manage this?"

"I think I got it. I stay inside this Embargo Program which will ensure nobody can eavesdrop. Then sign off and unplug. Take care of yourself, Mal."

Mal then took off for dinner at home while Lev logged in, applying Mal's current faculty access information.

Within thirty seconds he was able to download a fast revealing JPEG image of dubious resolution, one line of text encryption accompanied by an audio file which said, robotically, "command-shift-three," followed by the one line of garbled encryption – *noitnettaruoydeeniemsti* – that stayed on the screen for all of ten seconds.

Another audio indicator: "In 118 seconds this picture vanishes forever."

David instantly hit the "command-shift-three" just as the top few pixel lines of the image were beginning to reveal something. He stared at it. Time stopped. Then it all went "poof."

Chapter 41
A Classified Huddle

"So hi-tech it's classified," Jake grinned, paraphrasing his group's public face for interns; come spend "a day inside the imagination of the CIA." One of the few attempts at open lyricism in the Agency he'd ever heard.

But there it was. Untarnished atop the user-friendly info, online for anybody to read.

Claire Cosgrove Upton was now downloading a rapid-fire sequence of interesting images from her National Park for her partner, as she preferred to designate her husband of two years, father of their two girls, Jane and Marie Antoinette Cosgrove, identical twins, with a few very scientific exceptions. Both appeared to have early language acquisition, at 10 months; both had laughed by week three; both were now pulling paperbacks off their parents' bookshelves, and thumbing through them, including the ones – which were most of them – with no pictures.

The biggest difference either Claire or Jake could discern was that Jane liked to be by herself, and Marie Antoinette, perversely, liked to play in the kitchen, especially with knives.

"Do you really think you two made the right decision naming her Marie Antoinette?" Claire's mother asked, a few weeks after the difficult delivery. "I mean in retrospect, given her habits in the kitchen?"

"What the hell," Claire had said, quite freely. "I've read Marie's letters. She was totally innocent and got a really lousy deal under that guillotine. We intend to make things right."

It all made sense.

"Who is he?" Jake asked, examining the images as they came into focus.

"That bicyclist has taken twenty-two tourists back and forth into Poland in ten months. These images show us where he's sleeping at night."

"Under a tree?"

"Correct. Strange, for a tour guide, wouldn't you think? And look at this one."

Claire clicked a key on her computer and revealed a series of one-meter resolution images showing the young man cooking a meal on a camping stove that emitted no color.

41   A Classified Huddle

"Hydrogen. Unreadable. He is soooo hiding," she declared.

"From whom?" Jake asked, confusion in the air. "He's a tour guide, for God's sake."

"Day job. Night job – you tell me what you see."

Claire then revealed to her husband a series of images of the guide/camper that were clearly indicative of someone who knew the scientific research station well. A full 18.6411358 miles, 98,425.2 feet between the bicycle path and the research station.

"He moves with impunity, that's how confident he is."

"Confident of what?"

"That we're not watching him."

"Sweetheart, watching for what?"

Claire fiddled with one of her other mainframes, then clicked on a classified motion-sensitive video file and hit "play."

Both watched as the figure searched under woodpiles, then easily tricked open a lock on one of the doors of the research station and entered, where their surveillance ended.

"So he's spying on that research station? High finances surrounding mushrooms?"

"No. This guy was a grad student there two years ago, got his Ph.D., then started leading bicycle tours for barely enough money to make an open fire and roasted marshmallows for his dinner. This is the same guy, and the same scientific station that JV – John Vespers over at State – has been following for three years. We have that not only on the authority of your buddy in the Doughnut, but countless intel files from the Rio Summit terrorist attack."

"So the State Department is interested in business ties with a former Cold War nation? Way of the world, my dear."

"I presume you've read Vesper's MO?"

"Shady. Half-dozen affiliations. Several deaths on his watch. Clearly commerce-driven."

"Ruthlessly so. I'd love to send his wife a birthday present and a note: Let's have lunch. What do you think?"

"Really bad idea, if you're asking me."

Four days later, both Mr. and Mrs. Cosgrove received a fancy invitation to a luncheon honoring the visiting Prime Minister of Israel to Washington.

It was in their home mailbox in Fairfax, and had been sent through the normal U.S. postal service three days before.

"We do not respond," Jake reeled, that night at home.

"But isn't that rude?"

He whispered something to her, and then said, "I hear the kids crying."

"You do?"

"Definitely."

He led her to the girls' room, quietly opened the door to see them both sleeping in their respective beds. Then, with a "Shhh..." he brought her to the little girls' bathroom, and shut the door. He knew that if the Agency was watching them, he felt

41 A Classified Huddle

confident that they were above spying on the bathroom of two little girls. That would never look good before an intelligence sub-committee.

"Claire, you do realize," Jake continued to whisper, "every single frame of every file search you've ever done has been recorded, logged in and obviously scrutinized by several other sets of eyes. This is not your exclusive gig. I know it's your region. But that's only your job description. Not reality."

"Allan has never said much more than interesting, and stay on it. Nothing about the killings, let along the other snoops, the traffic."

"That just goes to my point. They are way ahead of us. That's how it works, usually. We just do our jobs, file our reports."

"So we should go to them with this? I have very mixed feelings about the assassin. Something tells me he's a good guy in over his head."

"You file a report and get it to Allan yesterday."

"Right."

Jake led the way back downstairs, grabbing a coat on the way.

Then, "We're out of cat food! Shit!"

"Remember: Friskies."

Jake drove to a nearby 711, went to the restroom and, once inside, at 9:25 p.m., called Eddie.

"Sorry to wake you."

"Jake, what's going on with you?"

"I need you to tell me if the Israelis are involved in this thing I first asked you about."

There was a long pause and Jake could hear Eddie holding down the phone as a young woman's voice restlessly asked Eddie "what time is it?" with a painful groan.

Then Eddie, walking somewhere in his home, finally answered, "Yes."

"Can you tell me, like anything?"

"Tough one. Remember that time I chipped a tooth playing squash?"

"Uh, yeah?"

"How are your teeth anyway? Listen, I better go. Take care, my friend."

Jake went into the 711 and purchased a bag of Friskies, putting it on a credit card, for the trace.

As he drove home the five minutes or so, it hit him: Eddie was referencing the movie "Marathon Man." He'd just given Jake everything he and Claire needed to make an informed bad decision.

Chapter 42
noitnettaruoydeeniemsti

It was Sunday morning. Lev had just had coffee and was reading the LA Times in bed – the obits, his latest fixation; the extinction – or not – of the Vu Quang ox, *Pseudoryx nghetinhensis*, found in the Annamite Range of Laos and Vietnam, one of the Earth's mammalian canaries in a mineshaft with respect to extinction rates, particularly in that part of the world; but, equally depressing, particularly for migratory birds, the slow death of the Salton Sea in southern California; a discussion on rising anti-Semitism… none of this very digestible and certainly cause for the glass of cold fresh orange juice that he was craving.

Sasha stopped him from heading to the blender.

"I wouldn't," she said.

"You're right," Lev conceded. He really couldn't handle sugar on top of the rash of bad news. His blood work had recently revealed kidney and liver issues.

"Stop all yellow corn and sudden dosages of carbs," his doctor warned him.

He went to his computer, sat down, and logged in remotely, using Mal's password to the site he had gotten into the previous day at the UCLA research library, one he'd started using four decades before. He'd seen a few changes in the way libraries did things.

Now he stared at the same incomprehensible gibberish: "noitnettaruoydeeniemsti."

"Sasha, I need to show you something."

She was drinking her morning coffee and came and sat down beside her husband.

"One of the few words that turns up absolutely nada on Google."

Instinctively, she said, "It's me. I need your attention."

"What?"

"Sorry?"

And then Lev himself got the utterly ridiculous reality staring at him. He'd grown up reading from right to left. It's as if somebody else knew that, not just Sasha.

At that precise moment he heard and saw his little annoying iPhone buzzing and vibrating, although he normally kept it silent. He often forgot to plug it in. There were few people left in his world that he cared to hear from. The world of David Lev had been fading fast.

Sasha, noticing the same, suggested that they were being watched.

"How?"

"You just logged in remotely using somebody else's password on a site you really know absolutely nothing about, correct?"

At 87 there were still plenty of things to do, and much to relish, or worry about, depending on the moment, but Lev was certainly not up to this. Enjoying the most impressive countenance of the spotted towhee in his backyard, the sound of a song sparrow, yes. Computer games or terrorist attacks, no.

His world had come down to mending aching bones, trying to control his blood pressure and pulse rate, and curb serious bouts of depression, the latter seemingly the through-story that nearly did him in.

She picked up the phone. "Hello?"

"Open sesame," came the order of a strange, fake-sounding robot.

She shut off the phone.

"What?" Lev complained, angry on behalf of the two of them.

"A voice said, 'Open sesame.'"

Chapter 43
Cro-Magnon

Beyond the word, written backwards, was a JPEG image.
"Do I open it, or not?"
"Who else could be watching?" Sasha posited aloud.
"Anybody, I suppose. Everybody."
"But it says it's '*private*,'" she added.
"Right. Fuck 'em."
He clicked to open whatever it was.
There was, within seconds, the strange wild figure. A wild man.
"It looks like some Cro-Magnon man," Sasha said.
The figure was bearded, scarred, impossibly remote from the modern world. The origins of humanity. Adam and Eve rolled into one mishap, by the look of it. Neither of them had the slightest idea what the picture meant, or why they had been chosen to look at it from a login at an allegedly private scientific chat room, of sorts, from a familiar library. Lev managed to copy and paste and e-mail the image to his iPhone.

Then, feeling sick, he went into the downstairs bathroom, his iPhone in hand. After peeing he had the strange idea to orient the iPhone screen so that the picture side faced the mirror above the sink.

He stared at the face that was now reversed, futzing with the various options for manipulating the image – Malcolm had taught him all this shit after the disastrous Power Point imagery at Rio – until he was able to gain greater outlines, coloring varieties, background, and luminance variations. Highlights instantly emerged that showed a face staring right back at whatever camera had taken the picture in the first place.

It was at that moment, Lev's own face, and that of the image on his iPhone as reflected in the mirror, that something strange and terrible began to converge; an inconceivable, indeed horrible recognition from the heart of the Rorschach. This inkblot was in color, in the thickness of a forest.

There, in the palm of his hand was the body, and the face of someone he knew from another century, another world. His own younger brother, Simon. Could it be?

However impossible a concept, there he was, staring back at him as only one's worst fears and nightmares do.

But also one's hopes.

Chapter 44
Machinations

He staggered into the kitchen, poured himself a large mug of ice water, then went into the living room and collapsed onto the couch.

Sasha had tried for years to teach her husband about the pleasure of rare books, of ice cream cones, of simple dates together – dancing, chamber music, a movie with popcorn – anything to take their minds off that which had so invaded their twentieth century, creeping well into the twenty-first.

But as she came into the living room, she immediately understood that her husband had plunged into some terrible mire. She knew it had to do with back then, and there would be nothing in her power that could wrest the demons from whatever had happened in him.

All her efforts over the years to make her husband at least pleasantly accepting of the world had been labored, but unsuccessful. He managed by his own secret means, never truly divulged, because… what was the point? Sasha was not as overwhelmed by the same melancholy, even though she lived with it 24 hours a day.

Her beloved was fixed in a dark place. She loved him all the more because he was innocent. He could not help what had happened.

***

Time had truly stood still. Crawled from some Ordovician swamp a half-billion years ago covered in slime – not fitted in 130,000-year-old white-eagle talon jewelry, as was the day's rumored paleontological findings from somewhere in Croatia where at least one tribe of Neanderthals apparently bedecked themselves.

No, this appeared to be a study in utter desperation. But was it? How could it have been sustained? A miasma of every unthinkable dimension in time, geography, the throes of the worst beastiality; a collision that could not be ecologically conceived, biologically birthed, culturally envisioned?

This single image of the so-called "Cro-Magnon" went beyond anything Lev had ever heard of, read about, or imagined in the annals of science.

He stared at the image, squinting, shifting, putting a magnifying glass to the picture and the blown-up rendition.

Sasha studied it with him. Silent. She didn't know yet.

There was nothing that could acquit David's shock and sense of total terror. A desolation made worse by the shock of it: that was *him*! His one, his only brother. The last trace of his family. His genes, backfiring in a manner that just made no sense.

"It's him, my brother," he uttered.

"Oh my God," she cried out, stifling her bewilderment in the light of a flash of illumination: the years that had transpired ... the vast distance ...

... Frostbite. Ticks. Pneumonia. Nazis. All those many decades. Not possible ...

He shut down his iPhone and then it slammed into him like the assailant slamming their SUV in Rio.

A cascade of exploding nerves, like a terrible mudslide that had commenced far, far above, beyond any visual or audible line on the mountain; no predicting or second-guessing the incoming disaster; a roaring descent of all that this or any life could possibly hold in store for but one person of aching blood, deeply diminished muscle tone and patched-together bones. A brain spinning into blindness. Thoughts detonating like a vacuous meteorite shower.

He barely had time to cry out.

"SASHA! HELP ME! PLEASE HURRY!"

She instantly gathered his pill containers and a bottle of water – "DRINK!" – and got him into the car, his seatbelt locked on, and sped off to the ER.

The doctors quickly got an ativan IV to course through his black-and-blue veins – he bruised easily, thin-skinned at this age – and it made him feel like his chest was caving in.

"I think I'm having a heart attack," he called out.

"No, you're not," the attendant nurse assured him.

With the help of a whopping dose of atenolol, they got his respiration down from 22 to 12 breaths per minute. A three-day respite was ordered so that Lev could be monitored closely.

As he lay there in Westwood, deeply disturbed and uncomfortable by the garish lights and noises at all hours of the day and night, the events of the summer of 2012 rushed over him with feverish and revolting memories. He never realized how bad a migraine could be. And then the café, and the tragedy that swept over his young colleague for no plausible reason. Lev would have traded his own leg in a heartbeat.

All of it a mere prelude to ... now.

Inevitably, minute upon minute, his mind fixated on the image, on two images, actually: they had been separated in a single frame. The first was the large full shot. Grainy but clear enough, the second a blow-up, which was far more grainy, but made it indisputable in Lev's mind. The seriously primitive-appearing individual could not have been anybody else, as much as David Lev hoped it might be. There was that uncanny genetic connection. Genomes. Chromosomes. Diploid deceptions.

Regardless of how many physical morphs, one sees through the outer layers. The images revealed David Lev's little brother. There was no doubting it, despite huge changes. The images had been taken sometime prior to the Rio+20 Summit.

That was the reason *techsupport* had contacted him in the first place, Lev pieced together: To present the spectral truth to the one person upon whose own destiny the sibling's fate apparently hinged. Now it all emerged in a kind of freakish clarity.

The thoughts began to unravel. The part of that equation ... I need your attention? Those notices on the bulletin board at Pavilion 3.

Lev was scared shitless. New levels of panic that transcended a mere attack or two. This was material injustice, suicide, every conceivable nightmare crammed into one instant, and delivered, no less, by a stranger with whom he was nearly blown to smithereens.

As he lay there, dying, not dying ... And who else was watching, reading, listening in? Clearly Taman Chernichevsky had to wait, nearly three years, to ensure that the eavesdropping would not prove disastrous. Or to check if, in fact, they were still watching, listening. Had they been? Who else had seen the imagery?

***

And before he knew it, those hours, that moment was upon him. In his house. Alone with his wife and one aged parrot. The other had died the year before, in David's lap. Euthanized as a result of old age, a brain tumor, neurological disorders proving, day by day, to render him incapable of enjoyment. Of being who he was. This is how it went.

Lev sat before a muted TV. Incapable of fixing himself a drink, his hands were shaking so badly. He put eight cubes of ice in a glass. Three others fell on the Spanish tile floor.

Then he realized it needed water. He could not hold his hands steady enough to do more than contemplate his need for a glass of water. He was suffering terribly. The whole house was coming down upon him. His universe had imploded. He was well beyond the time in one's life for a shrink. Antidepressants, definitely. But no: David Lev needed to think, and to act.

Sasha went and filled the glass with water.

"Do you want to talk about it?"

"Wait."

He proceeded to send a return text message – it had been four days before this response – requesting a phone number. Sasha could easily see the difficulty her husband was experiencing.

His hands were shaking.

"My God. I've never seen you tremble so."

"Why don't you use the land line?" she said. "Whoever it is."

"No."

"My dear one, you're shivering." She put her arms around him.

The number came back, not heeding whatever security concerns might remain at large. David Lev then tapped in the numbers on his cell phone, unconcerned with the 24-hour clock. Nothing mattered but the voice. And then, that voice answered.

"It's you?"

"Yes."

Both men were silent. Then, as David started to ask, the voice on the other end said, "Less than ten seconds to avoid a trace. Just come. Capital. Main train station. No flights. Text your arrival time. Dress warm. I'll be there." And he hung up.

That was it a kind of grotesque clarity.

Lev shut down his iPhone, took off his shoes, and lay back down on the bed, his eyes closed, his right arm across Sasha who hugged him silently.

She could feel every part of her husband's beating chest, heavy breath, the fear rising in him. She couldn't put anything together. Of course she couldn't.

"Sometimes, I dream about Ireland. How wonderful it was. How nice it would have been to have stayed there. To have given up all this," she said softly, like some girl with flaxen hair. "Or to remain in that holy place of Grieg's "'Lyric Pieces'," she mused, holding on to David, and aired aloud, like her evanescent memories of that vouchsafed melody from "Cradle Song."

© M.C. Tobias

Chapter 45
The Irish Connection

Sasha Darby-Lev's father played chess as a young man with a close friend who knew about strange things and always used to surprise Sasha with tidbits of the trivial encyclopedia in his brain. Like the best edible seaweeds in Ireland, and the origins of baking soda. Or the number of toilets in all of seventeenth-century France. This latter fact was a trick question, he said, which she never understood.

His name was Caleb Halevi "Collins" and he used to teach the history of the British Isles at various schools around Ireland and England, even Trinity College, briefly, before collapsing and dying on Stephen's Green one autumn afternoon during his most productive years, long after Sasha and David had moved to California.

Collins had specialized in topics pertaining to the monks of medieval Sceilg Mhichí, or, as some are wont to pronounce it, "Skellig Michael," that western-most pyramid in the southern Irish seas, enshrining the best of ancient Irish civilization, rookeries of razorbills, puffins, and other birds. Sasha had been to the pyramid and seen the beehive hamlet comprising six remaining huts on the south peak, wherein during the turn of the first millennium monks lived, prayed, and kept meticulous notes of their spiritual lives. Today, this UNESCO World Heritage Site was also a tourist haunt, weather permitting, and one of the most spectacular geographical intersections of ancient history, culture, and learnedness on Earth.

Sasha and her father had once had the opportunity to accompany Collins out there. It was drizzling, and the climb up the rocks was treacherous, but they made it – only to be stuck there in a downpour and have to spend the night in one of the huts.

Sasha's husband had not been impervious to its powers, either. David had obtained a scientific permit to visit the island long after they were married. He spent two entrancing weeks there observing the birds, and deeply aware of, and forever conflicted by, the fact that the man who first made it all possible, Collins, David's uncle, was also at the center of what could only be viewed in the aftermath as a calamitous, if soteriological bifurcation: David's separation forever from the world and the family he had known back in Eastern Europe.

Of course, that blood tie – Irish blood – had also saved his life and made his marriage to Sasha and their subsequent journey to America possible. Sasha had been born in London, the daughter of longstanding Jewish rare book dealers, among the first in that trade along Cecil Court.

Her stomping grounds as a child included the pigeon-enlivened expanse of Trafalgar Square. And she also came to know most of the famous paintings at the National Gallery next door. Her favorite was the famed "Arnolfini" by Jan Van Eyck. But she was also extraordinarily partial to the Gallery's "Wilton Diptych." She was mesmerized by its immutable beauty, the quintessence of spirituality.

But it was her love of the birds that brought her and David together – that and Collins. David was a young boy in Dublin when Sasha was living in London's undergrounds during the air raids. She vividly remembered the loneliness, despite being one refugee among thousands at night, and clinging to the books that were her family's traditions, including much of Dickens, Gilpin, Ruskin and their friends.

Her father had donated one of the earliest known Talmuds – the basic book of Judaic laws, regulations, customs, and all things Jewish – to a collection in Ireland that bought the family some valuable brownie points in life. But that was later on.

It was a few years before the bombings that the British Jewish Refugee Committee had enacted the Kindertransport program of some 10,000 Jewish orphans and/or children without their condemned parents, to Northern Ireland, many temporarily to a famed farming estate. Kids began pouring in one year prior to Hitler's invasion of Poland.

David Lev had managed to avoid the chaos of such an orphanage or the bombing raids of London. In fact, he had come straight to Dublin in 1937 at the age of nine, in the company of his (at least by his own family standards "Famous Professor") Uncle Caleb.

This kid from Eastern Europe could recite loads of Latin – not Church Latin, but "bird Latin," so to speak. *Crex crex*, for example, the Irish corncrake. Of course, the boy did not yet know that that crex actually made a sound like "krex"... nor that it would become an endangered species. But one of the Customs officials knew the crex because he'd heard it growing up in the countryside. He thought the newcomer's mention of it more than astonishing.

Like David's own father – who never quite understood why his older brother, Caleb, should feel so insecure as to adopt the Irish last name of Collins for appearances' sake – had never hesitated to encourage all the other trappings of learnedness in both his sons, David and Simon, although he always remained a village farmer. No pretensions. A deep and abiding love of his Hasidic upbringing, soon to be shattered.

As Lev later came to realize, if only every other member of his family had joined him that day. It all hinged, really, upon a single morning in question when everyone who could be there gathered at the nearby train station to offer their – they thought, temporary – farewells. When the air was still relatively light, despite his memory of the Polish cold. That white winter's light. The sum total of electromagnetic spectra, healthy, illuminating. But that was all mere embroidered thinking that emanated from a heated university office years later.

45   The Irish Connection

It was even colder to the north, his grandmother always said, referring to Vilna, or Vilnius. Half of his "tribe" were Lithuanian, from Vilnius and Ukmerge Uyezd, or Vilkomir, in the district of Kaunas Guerniya. The other half were Poles, from the region between Bialowieza, that primeval forest, and the swamps of the Biebrza River in Burzyn.

David's world, the world of the Levs and Halevis – for at least 49 of them – would have been entirely different had they all decided that morning at the train station that Ireland sounded better then northeastern Poland; even that little village just miles away, safeguarded under the Soviet tyranny of Stalin, might have proved safer.

But they did not leave on that train, nor walk across the border. Their village of fewer than 500 people, almost entirely Jewish, somewhere between the great ancient forest and the lumber town to its westernmost corner, held all the pillars of life that the extended Lev family thought meaningful.

It was a village of great joy, or at least as configured in the now bedeviled memory of Sasha's husband, who lay on their bed in a paroxysm of unknown depths, confusion, and a feeling that seemed to him an implacable harbinger of the end of his life. The first time he did not fear openly articulating the contemplation of suicide, with every rational reason for taking his own life.

***

Dublin had seen the construction of its earliest synagogue in 1660, and an elected Lord Mayor in 1874. By late 1937, just months after David had accompanied his Uncle Caleb to the thriving *entrepôt* that was Dublin, Jews were accorded representation in parliament under the Second Irish Constitution.

Moreover, there was even an Irish Chief Rebbe. When the War broke out, it was clear to Collins, who tried to explain the situation to David, that it would be a while before he would be reunited with his parents and his younger brother, but not to worry. He naïvely assumed that Churchill and the Allied Forces had everything under control with respect to the unstable psychopath currently filling in: Adolph Hitler.

In the very summer of 1937, the Rt. Hon. Alderman Alfred Byrne, Dublin's Mayor at the time, created the Jewish National Fund to show the nation's sympathies with the notion of a "national home" for the Jews in Palestine. For people like Caleb, there seemed to be an inevitable force for good combating the cumulative evil of the pogroms and the rise of the Third Reich. How different from the "First Reich," the Holy Roman Empire. He was, after all, a historian.

But nothing quite worked out that way.

Chapter 46
The Decision

"What happened?" Sasha finally got the courage to ask her husband. She now sat evenly on the side of the bed.

"I have to go Belarus."

Sasha moved across to the headboard, propped three pillows behind her, and took her husband's hand.

"Have you ever even been there?"

"Not since its independence. When I was a child, the border areas were murky, so I probably *was* there, at least on the forest fringes somewhere. I recall images."

She thought about it, then told him with clarity, "If you're up to it, you have to do it."

"Yes."

"It's dangerous, isn't it?"

"At my age, everything is dangerous."

"The same forest from near your village?"

"I think so. I have to do some research. It's the Belarusian side. There is every reason to doubt the veracity, every reason but that photograph."

"How will I be in touch with you?"

"I have no idea. These days Androids and iPhones tend to work in most places."

"What if you – well..."

"Disappear?"

"People disappear. You almost did. And what happens if you go and get yourself arrested? At your age?"

"There are American Embassies everywhere."

"Terrific. You saw how well that worked out in the American-controlled elevator in Rio. And what will they do if you find him?"

"That's the problem. What to do."

It was a hollow admission of both inexplicable guilt and the incredulity of imagining the scenario: What could he possibly do? It was an encounter that would build up in his mind in the coming days with horror-like provocations. He dreaded that forest now.

"What if you have panic attacks in the forest?"

"Sasha, he was killed in 1939. My whole family was murdered, along with the entire Jewish community. The village is now classified as 'vanished.' You know all this. Panic attacks? Are you kidding? I *have* to go."

Embers still glowing. Lev could not get the idea out of his mind. As ecologists, he and Sasha were stunned, but also in tune with those embers, with a remarkable dilemma, the spellbinding hope that some other family member had actually survived the War. But in what sort of mental condition? And could they get him out? Or, worst of all, might they not trigger an avalanche of media coverage and global fascination with some kind of freak, an unraveling that could only prove disastrous.

"This is a nightmare," Lev uttered, out of breath.

***

In the coming days and nights, David Lev spoke with no one but his wife, dropping a few obligations, focusing on nothing but that forest, the medical provisions he'd need – enough Xanax to sedate a walrus – and boning up on everything possible, including trains in and out of Minsk, visa requirements, and the vertebrates of Belarus.

He used Google Earth to the extent possible to better assess what was potentially in store for him this time of year, to adduce which birds would be there, which mammals. And, most formidably, how a human being could have survived in such a forest for almost an entire lifetime, alone, in constant hiding when the modern world had encircled that same forest. The proliferation of the human presence was obvious at a glance. Not to mention the eco-tours that were touted on the Internet.

He simply could not fathom actually meeting his brother. In what language would they communicate? His brother had grown up in a household that spoke mostly Yiddish, but also Hebrew, some Polish, Lithuanian and, as necessary, broken Russian. But that was David's memory of it. He actually could not truly put a voice to the little kid who was his younger brother by five years.

When David had migrated to Ireland, Simon Lev would have been four or five years old. David was going to be 88 in the coming fall. If the authorities were seeking out a wild child in Bialowieza, they surely wouldn't have to look too hard: just find the two oldest cripples in Eastern Europe clumsily tripping through impenetrable forest, crying wolf, crying for help. It had to be that easy.

A wild child, 82, 83 years old, still out there and presumably imagining that World War II was still going on.

The concept was of mythic dimensions. Everything about the science of it Lev found to be impossible. He could not imagine his little brother trapped in a warp of historic depths that had condemned him to this continuing tragedy – the living and re-living every single day and night of a colossal horror that he might well have witnessed.

A horror that David had managed to let rest. His silence about it had begun almost as soon as the intimations of what happened to his family had been elucidated.

46 The Decision

It had started five years after David had left his homeland and was well established at a private boy's school in Dublin. Thanks to his uncle, he quickly had the appalling opportunity of seeing it on the dinner table over a weekend, a small brochure whose cover read:

REPUBLIC OF POLAND

Ministry of Foreign Affairs

THE MASS EXTERMINATION of JEWS in GERMAN-OCCUPIED POLAND; NOTE

addressed to the Governments of the United Nations on December 10th, 1942, and other documents

Published on behalf of the Polish Ministry of Foreign Affairs by HUTCHINSON & CO. (Publishers) LTD.

London: New York: Melbourne

Price: Threepence Net

David now remembered that moment with inconsolable dismay.

Chapter 47
London

A January 3rd, 1942, meeting in Saint James's Palace in London had resulted in subsequent meetings and condemnations by everyone from President Roosevelt (who is said to have asked about horses, not Jews); by Winston Churchill; the Polish Prime Minister in Exile, Wladyslaw Sikorski; and Vyacheslav Molotov, the Soviet People's Commissar for Foreign Affairs, yet nothing had been done to stop the continued epic of hatred unleashed by the Third Reich.

The emissary for this mission was the highly decorated Catholic-raised diplomat, Jan Karski-Kozielski, who had begun his work in the Foreign Ministry at the beginning of 1939 and had come to London to report upon what he knew of the extermination of Jews in camps created by the Nazis in German-occupied Poland.

Of course, Molotov had secretly signed a Nazi-Soviet non-aggression pact with Joachim von Ribbentrop, in August of 1939. Nor could anyone have possibly anticipated the cool reception with which Allied Forces would ultimately receive the information.

In 1982 Karski was recognized by the State of Israel and its official Yad Vashem memorial as a member of the "Righteous Among the Nations." In 1994 he was nominated for the Nobel Peace Prize and died in the U.S. in 2000. On May 29, 2012, President Obama awarded Karski the Presidential Medal of Freedom. Accepting posthumously on Karski's behalf was a Holocaust survivor and former Foreign Minister of Poland.

Lev had never forgotten that little three-pence brochure of 16 pages or so, shown to him by his Uncle Caleb back in Dublin in 1942. Who could forget a phrase like "the mass extermination of Jews"?

The rest, various details about the remarkable Karski, Lev looked up on the Internet during the weeks prior to his journey to Minsk while his tourist visa was being processed, and his mind and will were saturated with whatever ammunition he could imbibe to prepare him for the unknowable.

He worried about that fellow from the U.S. Consulate who first visited him at the hospital in Rio, particularly in light of the alleged scientific breakthrough Taman had declared over the telephone. It reeked of international commerce. That was a near given.

Sasha helped to the extent that she could, as David found it difficult to focus on so much history in so little time, his fingers clumsily attempting efficiency on a computer keyboard and his heart pounding, when his whole life had been by and large carefree, easily lived outside the mutilations, and exhaustive and infuriating political intricacies that World War II had tapped into, eliciting every conceivable condemnation of human nature, but continuing on in a vast question mark of restraint and unsteadiness.

A sickness gripped Lev's brain, leaving him exhausted, nearly vanquished before he could get out of bed. An aged professor with really nothing to say, nothing to show for himself. A failure of a man against the backdrop of the worst evil ever perpetrated. He could not defend himself against the plight and truth of his own brother, two quintessential realities now looming for an older brother who had nothing of substance to offer a record of atrocities and the unthinkable within his own family circle.

He was, in essence, a doomed being, and said so to his wife. She could not exactly comfort him. She herself was too terrified, realizing that her husband was about to leave for the same Eastern European forest that had massacred every one of her in-laws – every one but Simon.

During those very weeks and months in southern California, and in Europe, a few organizations were struggling with various ideas about how most appropriately to enshrine the concentration camps. One group of architects took their cues from the Torah, contemplating beautiful forests, thinking over the quintessence of logs, the nature of nothingness, about how nature and humanity together would one day look back at the more than six million victims who went up in smoke, or were tossed into anonymous graves. Would they believe it could ever happen again? Would some consider the beauty of a forest to have embodied the best of humanity, obliterating the bad dreams that happened to be the actual bottom rung of humanity's evolutionary ascent from the mud?

Would such forests counter, in the minds of future generations, the degenerative, devolutionary horror that was clearly part of the human genome – the notion that a small piece of Hitler lived in everybody? That – as Jean-Paul Sartre put it – even if there were no Jews, there would be anti-Semitism?

Now, David Lev entered into the discussion from a perspective he could still only half imagine; a plinth of guesswork cantilevered by a single photograph alleged to be real by a man he had met just minutes before the two of them were nearly killed in an explosion.

It was not easy to grasp that essential conundrum: these two men met at the Rio Summit, and everything else that had followed was simply ghastly. Yet, there was also the question of destiny written large between events, continents, centuries. No escaping that, so the thought of boarding a train in London bound for Minsk indoctrinated David with a fear only half phrased, scarcely realized. Every detail was the working out of a clumsy maneuver by a man without options or any kind of future. He had to do this, even if it was about to kill him.

47 London

The panic attacks, increasing now, and the pills he had to take to control them, only furthered the dizziness of this undertaking. That and serious arthritis in nearly every bone in his body, which seemed to be getting worse by the minute. From a lifetime of field research around the world in harsh places, chronic insomnia, wine, and pills. Now, psychosomatic nausea, and the incessant pain in his left shoulder from the bullet in the café.

But all such trepidations were of no consequence. The same physically excessive compulsions of his scientific career now defined an arc of the inevitable with zero alternatives: a brazen resolve to find out the truth about his brother – he had no other choice. One does not turn his back on such revelations – and somehow work out a resolution for the two of them that he might, or they might, be able to live with, or not to live with.

"It's just two and a half days out of London," Sasha tried to comfort her husband.

She had booked everything. The Eurostar from St. Pancras to Brussels, then on to Cologne by ICE high-speed train, then overnight on a Polish sleeper-car, the Jan Kiepura from Warsaw Centralna. He would arrive in Minsk on the morning of Day 3. The weather would be dreadfully cold, breaking all kinds of awful records, she reported.

Lev called Taman and informed him of his details. And that was the extent of it. A normal Tumi suitcase, with a backpack folded up inside. Clothes and meds for one month, although he very much hoped his trip would be half that long. Of course he didn't know what to expect, other than the fact, according to Taman, that there would be no Wi-Fi at the train station, nor that he should even think about taking out his iPhone in public.

Then the day arrived. Sasha, still a more-than-adequate driver at 85, aside from her habit of never truly ascertaining the extent of the blind spots in their eleven-year-old Mercedes 450, drove her husband to LAX, fully utilizing their "limp along."

"I love you. Be careful," Sasha whispered into his ear.

He took her in his arms, looked at her, and responded in kind. "Take care of yourself. I love you more."

"I know," she pleaded.

It hurt. The moment stung for both of them. A lifetime, right there, at curbside. Totally insane.

They hated partings at airports. Usually Lev took a taxi rather than imposing the drive to LAX on his wife. This time, things were different.

Chapter 48
A Glazed Twist

"Tit for tat," said the – at first – unrecognizable voice.

"Who is this?" Jake Cosgrove asked, sleepily, as he saw from his alarm clock that is was not quite 4 a.m.

"Go where you can talk. It's me." And the phone went silent.

"What?" Claire said, sitting up rapidly in bed, sounding so groggy it was as if by rote. She was basically still asleep.

"You mentioned you were out of tampons."

He was nodding compassionately.

"Oh fuck, you're so right."

"And I guess you really need 'em now, at this ungodly hour?"

"When you need 'em, you need 'em. You wouldn't know."

"Right."

Jake got dressed and drove to a nearby 24-hour convenience store.

He purchased not only Tampax, but condoms.

"Where's your restroom?"

The man angled his unruly face towards a rear corner.

Jake locked himself in, pulled out his G1 cell and called Eddie back.

"What's up?"

"This appears to be about two things."

Eddie was speaking from an underground somewhere in London. The noise was loud, unmistakable.

"OK?"

"A mushroom and survivors."

"Survivors of a mushroom?"

"They can be very poisonous."

"Right?"

Jake was waiting. He could hear that Eddie was reaching a tube stop.

"I have to step out now. Poisonous, yes. And very, very profitable. According to not Henry."

"Not Henry?"

"More like Titus Andronicus. Oh, one more thing. A lot of people are still alive, and they have everything to lose. The game is in play and it would appear there is a witness. How are the twins?"

"Great. You, I mean how's your family?"

"Take care of yourself, mate."

Chapter 49
Warszawa Centralna

Lev was already exhausted. Just spending several hours in downtown Warsaw at the Dworzec Warszawa Centralna train station had drained him. His first time back to Poland since childhood. He nervously ate a sandwich and had an espresso, taking in the busying cramp of heavily-wrapped Poles. Terrifying déjà-vu. Memories that he had long suppressed.

Even the odor was freakily familiar.

Then back on a train, and a ten-hour journey. Beyond exhausted. I can't do this... he kept saying to himself.

The forests, marshes, and fields were still languishing in the extraordinarily deep freeze of winter as the train sped noisily through an alien countryside. Climate change seemed to have actually exacerbated Poland's long-standing meteorological brutality, which Lev remembered with anything but fondness. A child cannot forget the terror of being cold. Of one's toes going numb, indoors. Of actually discovering a frozen rat beside the outhouse. A solid block of black ice that had been a rat. How does that happen? Frozen whiskers. Frozen urine that had leaked from the mammal as it suddenly went into whatever biological state of instant freeze had overwhelmed it. This was an image in his mind from childhood that Lev still could not shake.

The sunlight was filtered through cold racing mists. He looked sheepishly at fellow travelers, a dull ache that ruminated upon the various misfits of life and those who had arrived at some level of equilibrium. The quantum mechanics of evolution had become all the rage in biological circles – how a single cluster of molecules might mean a scent, or a taste, at a 2,000-foot altitude to a migrating Arctic tern, for example.

But these travelers seated on all sides – who were they? What worlds or longed-for ideals did they inhabit?

Lev had obtained a tourist visa within 12 days, no questions asked. That was a surprise, as he'd expected the worst. He wrote down that he was going to visit the famous library, and perhaps do some bird-watching in the springtime. That was the extent of his declaration.

He had filled out two cards, one of which was taken by Belarusian guards upon reaching the border area. The formalities were efficient and without incident. No different than having a ticket stamped by officious and systematic authorities on every French TGV. The ones you could never fool. The other card he stashed away in his passport holder and would make damn sure he had with him upon leaving the country. Leaving. The idea of getting away, getting out was already weighing palpably upon him.

The deepest-seated fear that he had entered a labyrinth from which there was no escape.

Escaping all this: That prospect suddenly faded away, now, as he recognized the much older-appearing Taman waiting for him quietly, standing near a steel pillar inside the station, nervously smoking a cigarette amid the hordes of weary travelers who had journeyed, like Lev, throughout the noisy, smoked-filled night from Warsaw.

Chapter 50
Pryvakzaĺnaja plošča, 9:07 am. Minsk

Taman put out his cigarette in a waste disposal. One didn't dare stamp it out with a boot. Not in Belarus. The two men shook hands lightly.

"How was your trip?"

"Yeah. Fine. Thanks. You look older."

"And you do, too. Best to take a taxi. Let me help you with that."

"It rolls. I got it."

"Please."

Taman took Lev's suitcase and they easily hailed a cab. Most people used the Metro, one of the finest in Eastern Europe.

They had previously agreed by phone that David would stay at Taman's apartment. The Chernichevskys had two guest bedrooms, one for each of their kids whenever they visited.

"How sizeable is Minsk, what, about 2 million people?" asked David with a routine nonchalance.

"More. With nearly 140 libraries, lots of universities, and of course churches. A lot of museums."

"How many cemeteries?"

"I don't know. Plenty."

"What about the number of remaining Jews?"

"Good question. I'm going to say maybe a few thousand. And zero-point-five, half of one, if you include myself."

This got a reflex rise from Lev's eyebrows. The professor couldn't help but realize that his host now looked him straight in the eye, a kind of physical truth he had never noticed before. This "host" in Belarus beside whom Lev had nearly died.

A man who then mysteriously disappeared amidst a chaotic inferno, in the middle of the night, then just as suddenly re-emerged three years later to commandeer Lev's otherwise fully-deserving and tranquil retirement. No such luck.

Continued Taman, "We have at least four synagogues."

"Which half?"

"Sorry?"

"Which half is Jewish?"
"My grandfather."
"Less than half." David smiled. "But who's counting?"
"And there you have our Rhombi."
"Vy chočacie, kab spynić?" the driver asked.
"Nyet, nyet," Taman answered in Russian. "I presume you'd like to get home and rest?" Taman asked Lev.
"Yes. My goodness, so that's your famous National Library?"
"The Rhombi. Yes."
"Quite impressive."

Within ten minutes they were in front of Taman's apartment complex. That plateau upon which most Europeans, Western, Eastern, actually lived, worked and dreamed. Of this struggling, middle-aged mycologist, half Catholic, part Jew, who found himself in a country still perceived by many as the last dictatorship in Europe, even if it was less anti-Semitic, xenophobic, more just, probably far more interesting than many surrounding nations.

They took an elevator up two floors and then turned left into a hallway that spilled in generous dimensions – sunroof, even Zen garden elements in the open courtyard below – towards three flats. The last was Taman's.

Genealogy, family, denial: each of these three components of the life we tend to call our own, had been superseded by a monstrous priority; the invocation of an unwanted piety; the remorse that comes from unchosen siblings and parents, circumstances, and the size of a baby crib, the challenge of a lasting open heart.

And Taman Cherneshevsky himself, the man who was singularly responsible for this invasion of privacy, the breaking of every rule of decorum known to human life, now proven to be mere sophistry for those slaves of family histories.

"I'd like you to meet my wife, Sofia."
"Meaning 'wisdom'."
She smiled. "I've heard much about you," she said in excellent English. She took his hand with both of hers.
"I don't know that there is much to say. I hardly know your husband. But you know all about that."
"Taman has told me a little about your scientific work. You are famous."
"Hopefully not in Belarus," Lev said.
"Let me show you your room, Professor."
"It's David."
"All right, David. You've got Alexander's room, the older of our two. He has graduated from the rock music posters on the walls to biochemistry, thank God."
"At Moscow University," Taman added.
"Terrific."
"Moscow. Mixed bag, of course. We don't know where he is getting money for caviar. A bit of a concern, I should say."

Taman took the Professor's suitcase and showed him to the room.
"It's only for one night. I think we should drive out in the morning," Taman proposed. Of course, he'd already conceived of every detail of some plan.

There was an *en suite* bathroom, and as Lev started to unpack, Sofia came to the entrance to the room and asked their guest if he would like some fresh croissants.

"We have our own baker, so to speak. Half a block away. He's a trusted friend. Actually, a distant relative."

"I've never turned down fresh croissants. Chocolate, warmed for twenty seconds in a microwave, if that's possible."

"Of course," Sofia nudged her husband. "How do you take your coffee?" she added.

"Non-dairy creamer?"

"What is that?"

"Forget it. Just an artificial sugar works. Equal?"

She and her husband spoke rapidly in Belarusian and she went into the kitchen to put on hot water while he called his relation to put in the order.

It was a large duplex within a cheerful, Bauhaus-like façade containing two dozen such identical domiciles with bright splashes of color. Nothing like some of the Stalin-era complexes he had seen in pictures on the Internet.

There was an odor in the air that smelled unmistakably of the forest, or regions of the northwest around Portland, he thought.

He had read that Minsk was surrounded by thick woods, and now the brisk morning air, bright sun punctuated by dramatically lit clouds seemed to excite the many atmospheric scents of that geography.

The drive from the train station had revealed numerous parks. Aside from some monolithic remnants of the Soviet past, Minsk appeared to be a truly beautiful city.

Taman opened the door. The buzzer had rung.

A young man in an apron had walked across the street to deliver the order, Lev observed, retreating to the kitchen, where he helped Sofia with the coffee mugs and the two of them joined Taman.

Seated around the laminated red breakfast table, eating their croissants, drinking coffee, Sofia discovered a note jumbled up in one of the six croissants. It had been shoved into it like a fortune in a cookie. Sofia had been the one to open it.

She read the one line, then handed it to her husband, with a look of deep frustration.

Taman read the note, took a lengthy solemn moment and sighed. He then ripped up the piece of paper and went and tossed the pieces down a toilet, then flushed it.

"What?" Lev asked.

"Shhh!" Taman gestured.

"So no clapping of hands?" Lev asked matter-of-factly.

Taman glanced at his wife, who stood up and went back into the kitchen. It was more than she could handle at that point. Clearly, from Lev's perspective, other events had occurred in this household after Rio.

Lev started to speak, then – with another look from Taman – thought better of saying anything.

Taman went and turned on a classical music channel, at a high volume.

A Vivaldi concerto for two trumpets and strings.

"Nice speakers," Lev said. "Bose, aren't they?"

Taman wrote down on a piece of paper: We should leave as soon as possible. Twenty minutes. Something's amiss.

Lev read the note and then nodded at Taman with the surreal sense of being suddenly pursued by unknown enemies.

Taman moved quickly into the kitchen and shared a few whispered words with his wife, who in turn declared at an unnaturally high decibel level, "Of course, the Star Café for lunch, prior to the city tour, is a great idea."

Lev admired her agile complicity. She had soft brown eyes that exemplified the old Russian adage "more sinned against than sinned." If there was any consolation in rebellion against the old Stalinist Big Brother syndrome, it was absent from her bright gaze. Everything about her face defied the word "haggard."

She seemed unaffected, in her beauty, by that otherwise renowned fatalism of multiple generations that had successively been defeated, under every Eastern European sleight-of-historical hand. Yet, here she was, full-force, undiminished, buoyant, even bold, Lev recognized.

"I've been dying to see your famed National Library," Lev said. "I'm told it houses one of the world's greatest collections of old manuscripts. I wish I lived here. What a glorious city!"

"No doubt." And managed an audible whisper, "You don't need to overdue it."

Lev insisted on finishing a second croissant, brashly finished his mug of coffee, thanked Sofia, and went to their son's room to "unpack."

"You'll find plenty of hangers in the closet," Taman exclaimed in a welcoming voice, fully manufactured at this point.

The tension in the duplex had just gotten ramped up several notches. Lev wondered why, all of a sudden, if authorities had been keeping tabs on Taman. To what end? He was so tired that he really didn't give a shit.

What could anybody do to them that Lev himself had not already inflicted by his decision to make this trip? The strange eddies throughout the post-Rio catastrophe had left him perpetually on edge, almost nauseated.

Taman walked into the bedroom and whispered to Lev, "Just bring your parka, hat, gloves, and hiking boots. You brought your boots, yes?"

"Of course."

"And don't forget wallet and passport. No luggage."

"And don't forget wallet and passport. Leave any unnecessary luggage in that closet.".

Chapter 51
Vulitsa Kirava 13

"You didn't leave me time to bring an extra pair of socks, let alone binoculars, sweater, so many things. What are we doing? It's all smacking of an irrevocable malaise, by turns stricken and incomprehensible, at least to me. You?"

"Put it this way: Heisenberg Uncertainties. Much to do, no time to lose."

They had arrived at a fancy hotel. Ordered two suites. Taman had left his old Volvo with the valet. Lev paid in U.S. dollars, although Taman had not yet explained to him why it was necessary to pay three nights' worth.

Taman and the professor were given their room keys, radio frequency hotel smart cards – designed in Sweden, manufactured in Taiwan.

Within ten minutes the two men had reconvened and were quietly having coffee at the hotel's Star Café. The rooms were a mutually determined ruse.

"I know you were worried about my leaving the vehicle," Taman said.

"Easy target to bug."

"Exactly."

"And you think they won't suspect – three nights, empty rooms? I mean, assuming you are right about all this?"

Taman gave a quick and knowing shrug, then picked up his Android, dialed a number, and spoke ultra-softly in Russian. It was apparent to Lev, who eyed others in the restaurant, before turning to hail their waitress, that Taman was conveying some very explicit instructions of some kind. The machinations, he realized, had only begun.

Completing his call, he then asked Lev, "What would you like?" translating an order of additional coffee and whatever finger sandwiches the young woman might recommend.

Once served, Taman excused himself – "just going to use the loo" – went into the men's room, removed from his woolen trousers a stick of chewing gum, put it in his mouth, chewed, then went to the furthest sink, placed the numbered valet stub for his Volvo on the gum and – confident he was alone – adhered it beneath the sink, testing it for assured staying power.

He walked back to their table. "Shall we?" Taman said, paying their bill with cash.

"Everything," he said to Lev, "has suddenly gotten about a third more expensive. It's the oil and gas crisis. Welcome to Belarus."

They moved casually through the entranceway side doors.

"Where are we going?" Lev enquired.

"First, fifteen-minute stroll to our Church of Saint Simon and Saint Helena. Nazis once occupied it. Then it became a Soviet cinema."

"I have been to other churches devoted to Saint Simon. The coincidence is not a little troubling, wouldn't you agree?"

"To throw them off. Good Catholics enjoy sanctuary, an excuse to re-group inside an inviolable space."

"Then what?"

"We'll go to the Ploschad Lenina Metro Station."

"Right. Metro to where?"

"Just try and pretend to be a tourist for the time being. The questions are bigger ones, and they are coming, as you know."

Chapter 52
Minsk Passazhirsky

They sat in coach en route to Kamenets and on to Kameniuky Village. A metallic rhythm, the strident noise of which competed with their quiet discussion over the course of a journey 380 kilometers towards the southwest.

Lev studied the bare farmlands, the fragmentary patches of forest, mixed townships gathering dust on their way towards the fringes, where villages overtook the natural in one guise or another; these hamlets given to that nostalgia with their cows and horses fully exposed, as the train sped past a world of bygone miasmas.

Lev's mind converged more and more uncomfortably upon the stark exoskeleton of landscapes and inner demons inhabiting those vistas; horizons closing in at high speed.

Taman picked up his Android. Lev observed the crucial gestures, divining some scheme that had burst into motion.

At the end of the conversation, shutting off his phone, Taman informed Lev that a close friend had, as conceived, found the wad of chewing gum, then with stub, cash, and the wherewithal to secure essential duplicity, driven Taman's car from the hotel and was en route to a northerly location of utter insignificance, as far from the point in question as possible, within Belarus.

And the most crucial giveaway: someone was following on the highway.

Taman's heightened fears now translated into every suspicion. There appeared to be a staring man at the rear of their 2nd class cabin, something about the way he was adjusting his glasses and newspaper.

A compulsively imaged series of deceptions. The slowing down of the train. A change in the weather, for the worse. That man three rows over, with a graying beard, balding, a tough guy, working on his iPad, kept frowning over toward them. The ticket official pacing the coach car. Pacing too many times, Taman decided. It added to the mental wreckage he was tangibly calculating.

"What is it?" Lev asked. Taman was not proficient at hiding his anxieties.

He all but whispered, a linguistic device not easily disguised. "We are nearly ten million people in this country. One in eight works for the police or military, the highest in the world, maybe with the exception of North Korea. I don't know."

Since the explosion, notwithstanding his wife's insistence that he was being overly paranoid, there was every reason to expect the worst from just such a mole, or policeman; an inside job from the Academy; some kind of pernicious ties to pharmaceutical giants in France, England, the U.S., competitors at the Academy. A number of suspects had already arisen in his mind; he had kept tabs regularly, but could not account for this or that behavioral anomaly.

"You're frightened," Lev noted, staring sadly at the figure in deep turmoil before him.

"Yes. Wrong places, wrong times, as you say."

In their scant communications, to date, Lev had easily figured it all out, he thought. Taman was no actor, able to disguise his true feelings. The sensation of being on a live wire, or touching a two-thousand volt predator fence was real. It showed in his every look and gesture.

"Do you get headaches – since that night, I mean?"

"Yeah. I take Sumatripan. You know it?"

"No."

"Here." Lev took one of the sealed pills out of his bag. "It usually takes five minutes. Wonder pill."

"I'm going to buy a bottle of water in the dining car. Back in a few minutes," Taman said.

As he made his way three cars down, his head was throbbing massively. He had not managed well under the stress of what, he well knew, was his own initial complicity. He had exploited the forest for his own riches, so to speak.

The conquest of fungi might be different than, say, climbing a new route to the summit of some Himalayan giant, heretofore deemed impossible, but indeed conquest was not the inappropriate metaphor: a scientific discovery is always fraught with ego, inevitable jealousies, and the uncertain schemes, players, and motives attendant upon commercialization at any price.

In his case, the prospects were particularly harrowing as he had come upon not one, but two enormous secrets, their conjunction nothing less then alchemical, but more easily interpreted to be sitting precisely atop an ethnic, legal, and moral time bomb, not just possible gold.

A new cancer-fighting property, or chemical that could be applied in any number of highly lucrative medical arenas, possibly canned foods, fermentation, aphrodisiacs for those willing to believe in such nonsense, and herbal cure-alls. This boded serious dollars. Especially the so-called "immortality factor" part of it, or that's what the corporate goons were likely to imagine they could sell their marketing divisions, especially in Europe, where American companies were flocking off-shore for much-debated tax reasons.

The problem Taman was facing: It might be true.

Chapter 53
No Birds

As Taman paid for the water and headed back toward his seat, the train came to a razor-sharp, metallic-whistling, high-pitched halt.

"*Kvitki, pašparty, dakumienty, kali laska,*" the conductor's voice rang out.

Taman made it back to his seat just before two of the State Police came through their cabin, checking every passenger's ticket and ID.

"Your passport, David. This is routine."

"*Amierykanskaja?*" the policeman asked Lev.

"Yes. You're American," Taman said, masterly concealing, he hoped, his terror.

"Yes. From California."

"California?" the policeman said, delighted by the concept.

"*Jon z vami?*" the other policeman asked Taman.

"*Tak, my kaliehi.*" "I explained that we're colleagues, traveling together," he told Lev.

"Lakers!" the policeman said, or asked, but clearly a source of pride, notwithstanding thousands of miles separating the cop from those tall guys. "Best team!"

"Lakers very good," Lev granted, with a smile.

The policeman asked Taman what the American was doing here and Taman told him sightseeing. Both policemen laughed, then the Lakers fan, studying Lev's passport, said, "Deyveed Leev."

"Yes, David Lev."

"Jew?"

"Yes, I am Jewish." Lev held back his unspeakable protest. A rage within a panic.

"Why not stay California?"

"Bird-watching."

"*Sumna praĺnaja? Što jon kaža pra?*" the policeman asked Taman, with a most perplexed look, still holding onto Lev's passport.

"*Nablyudeniye za ptitsami,*" Taman clarified in Russian. Not *bored* washing, but *bird* watching.

Both policemen looked at one another, clearly perplexed.

"No birds," the Lakers fan tried to explain.

"Maybe I'll get lucky," Lev said.

"Here," the policeman said, placing a clipboard with just a few names written on it before the professor, pointing to the line where he was to sign his name.

Lev looked to Taman who showed zero emotion and sat motionless.

His hand was visibly shaking when he took the pen offered by the policeman and signed his name.

The policeman returned his passport.

"You speak good English," Lev said.

"No, but Lakers very good."

The officials moved on to the next compartment.

"What was that all about?" Lev muttered.

"Welcome to Belarus," Taman said, in the manner of a ventriloquist, as if holding a sewing needle between his teeth. "They know we're here. That's for sure."

As the train began moving again, Taman and Lev could see that the two policemen had gotten off at the station, along with what appeared to be a man and his wife, hobbled by three large suitcases between them, who seemed to be at the center of some confusion matrix.

Chapter 54
Unique Circumstances

Taman had never divulged the most pressing anxiety sitting heavily on his heart, his thoughts at night, the wide-open apparition that was both urgent and mind-boggling. A preconceived order of DNA, embodied by a surviving Jew; surviving beyond all conceivable probability as the explicit result of some amazing constellation of co-symbiotic relationships out in that forest.

Far too complicated to utter, even to a world-famous scientist implicated for no fault of his own, on a speeding train in winter. But he had made clear the case, and Lev was there to bring closure upon the unspeakable, by whatever manner might be appropriate, or possible.

Declared Lev, "You disappeared after the attack."

"I had to get out of the city."

"Who was responsible? Why?"

"It's a mess. Taking in more than 75 years of time. People with much to lose. A security guard, the close friend of my cousin, is the one who watched over you when you were driven to safety. I heard what happened, although I was never sure."

"A security guard? There was very little security. Somebody on the inside at the consulate, *my* consulate!"

"There was a friend of my cousin's who protected you on the way to the Consulate. But once inside… I never heard anything."

"Blood and guts. I don't trust anyone."

"They interrogated me, you bet your ass. Questions. I had no answers. Then it was over."

Taman sat quietly, trying to keep his hands from shaking.

Asked Lev, "So when were you last at the location?"

Taman steadied his response, knowing it would lead to a very complicated, even self-incriminating admission. He had given this moment much thought, trying to find the way to characterize what appeared to have happened.

"Two months ago."

"How much snow?"

"Deep snow, yes. Six, seven feet, I imagine. Not like this."

Lev stared out the window, passing landscapes trapped in the moribund. Not one ounce of anything smacking of romantic beauty, impressionism, only desolation. And very cold.

"And how long since you've seen him?"

"David, I saw him two months ago."

"Why didn't you tell me before now?"

"I was afraid. Spies."

"What was he doing?"

"Waiting."

"Waiting? For what?"

"For what, or for whom? In either case, I have no idea."

"I don't understand what you're saying. Where was he?"

"Up in a tree."

"A tree?"

"All I can tell you is that that tree is the reason they missed him. It was simply not thinkable, I imagine. If they'd had dogs, it might have gone badly. Even drones would miss it. Although there is rarely a drone and in this weather, no. Again, I am merely guessing."

"Up in a tree? How could an 82-year-old be up in a tree, in the middle of winter? I just don't buy it."

"Yes. A tree. But I suspect he might be using that as a back-up or additional food source, epiphytes, lichens, that sort of thing. His primary residence, if you will, is more likely a giant tunnel within a fallen tree. The interior of deadwood gives off heat, enough heat, I imagine, to warm a person throughout the bitter cold of winter."

"What did you mean, *they*?"

"David, others are searching for him. You must realize that. I told you this was urgent, from the beginning. I was not exaggerating. I thought I'd explained the others."

"They? How many others?"

Taman weighed his words. Then, "I'm not sure. The footprints were everywhere, but it can be confusing. I am no tracker. I know most vertebrate prints in the snow in that forest but this was complicated by multiple snowstorms, frozen rain, the timing. The worst winter, actually, ever documented. I did not see it. And I cannot trust anyone, not even Ulyana."

David threw him a glance.

"You'll meet her. Kind of a ranch hand."

"An old man cannot live in a tree in the winter, not in Poland, not in Belarus, not anywhere. Yes, of course, on a cozy warm boat in the Caribbean, fishing leisurely. That's all well and good if you're Hemingway's "old man and the sea" purveying pearly nautilus shells. How naïve do you take me for? Maybe you think you saw him, but it was a bear. A tanuki, I don't know. It's simply not possible, in my opinion," Lev uttered, frustrated beyond belief. Yet, he had recognized the photos. It was true

"I last saw a brown bear in 2010, a freak occurrence these days, and it was more in the northern reaches. The Academy will be working to re-introduce several of them from the Carpathians each year. It's a new program. The Nazis had killed all the bears," Taman explained.

"Fuckers killed everybody," Lev mumbled, with a hiss for a coda.

"Yes. Of course. I know... This was no bear. And certainly not the tanuki."

"Whatever."

"It was a man up there. The same man you've now identified."

Now Taman spoke with near melancholy. "There is something else I must tell you. I have waited to do so because, well, I wasn't quite sure how to tell you, David, or when."

The Professor just dug his brows into him. "All right? When is now."

Taman drew in a breath, then began, "I saw him a third time. I have a single photo I managed to capture."

Lev was perplexed, verging on anger without a direct object. He was feeling totally manipulated. Everything about this entire saga, a double-cross somehow. No particular perspicuity was any longer required to grasp that he was being played, or that was his impression by this time.

"Show me."

He removed the evidence from his briefcase, this time on the size of A4 paper. Glossy. Blurred but harrowing enough.

"I don't understand. Just the one image?"

"And, well a few others from two years ago. I'm sorry. It was too difficult at the time to explain everything to you."

Lev took hold of the image before him, far too frazzled to drill Taman about timelines or whatever chicaneries had predisposed this man before him to his own devices. "It's blurred. Why so blurred?"

"It was instantaneous, actually quite lucky. I did not intend to see what I saw. I'm sorry."

"You don't need to be sorry."

Lev studied it in every possible manner, but in the end it was what it was and no amount of the uncanny gaze could change the implacable temptation: his brother was there, really there, one eye, old, and appearing quite sickly.

Moreover, he was not alone. There he was, cloaked like an ancient monk, or homeless victim of African famine, hunched and apparently feeding, browsing, grazing – whatever one preferred to call it – alongside a half-dozen wisent, the most charismatic Lascaux cave painting-like celestial beasts in all of Europe.

Nothing could be more astonishing: the deep snow, the clawing at anything to eat, all of them, a group of like-minded mammals, and there was Simon, his little brother in the midst. The faint whirl of hot breaths visible in the image, coming from the giant Bialowieza bison whose own near extinction and subsequent revival, however nominal, was a tale about which much had been described in the literature of science. But not this. Nothing like this. It was an impossible image.

"These are unique circumstances," Taman reflected out loud.

"Unique?"

"I know. I've been doing field research out there for several years. I am aware of it better than almost anyone. Almost. Except for you."

"And what you are suggesting is highly unlikely. Biologically speaking. I know of only one alleged instance, in the northeastern Himalayas, but never proven."

"I'm sure you have also seen that famous American photographer's work of the Indians wearing animal skins. Rituals. Or for hunting, I don't know. But warmth, no question."

"Photographer? Which photographer?"

"Curtis. Edward, I think."

"Yes. I know who you're talking about. That's totally different."

"It's no different than Inuit. Eskimos, Laplanders, certain remaining Siberian tribes. Boreal forest peoples who dressed in animal skins to stay warm. It is basic."

"I'm well aware of all that. The difference is, they are not solitaires. They have communities, they hunt, they have fires, huts, igloos. Women. They live in vibrant communities."

"He doesn't know the War ended. He's still hiding. There is no other explanation. And he is hiding with tremendous proficiency because the obstacles to doing so are enormous. A lot of foot traffic out there. And surveillance."

"But so many years have gone by."

"Of course. But there are two emergencies in one."

"Guilty people, monsters, still alive. One rich oligarch could snap his fingers. Your brother is a target. He's old. Possibly sick. He was hunched over. Feeding desperately, I believe. You can see it, right there."

Lev was still holding the image. He focused squarely upon his brother.

Continued Taman, "The guilty will take him out. That's the expression, is it not? I have seen the odd bullet hole in an oak tree. My guess is, someone, or several people, from the looks of it, fear your brother might be able to identify living Nazis responsible, and their Polish or Belarusian collaborators. And the very fact that he has survived all this time. He's getting his protein, his variety of food sources, somehow, and I believe I have cracked the mystery of that as well. Indeed, it has reached the point of an equally exposed crisis. I may be on to something that is truly, how do you say it – in French, I think, *'changement de jeu'*."

"Game-changing."

"Yes. Precisely. We say *huínia mianiajecca*. It has been a focal point of my research for years, by total coincidence. A genus of fungi. You understand? There are multiple problems. They all come down to this situation: your brother represents a colossal set of issues – political, historical, scientific. It explains everything, beginning with the bomb blast."

"But in a National Park, in the twenty-first century?" Lev incited. "Are you fucking kidding me? There has never been a bona fide case of a feral child. Not in France, not in India, not in Africa. And certainly not like this. I know that forest from childhood. It is ruthless. And he is no child, not anymore."

Lev's mind was wildly tracing through what little literature he knew on the topic. Over three-dozen famous wild-child cases from fourteenth-century Germany to twenty-first-century Los Angeles, the latter being a criminal case of grotesque incarceration.

Said Taman, "There was that Japanese soldier, lost in the jungles near Manila."

"I do remember. But this is not the Philippine tropics. I well remember a lot from my greatly dwarfed childhood. The winters were fierce. My mother used to hurry us indoors as the blizzards approached. *Beproti staiga pūga*. I remember that. Like *katabatics* that come in the Antarctic within seconds. Hundred mile per hour winds. One is quickly engulfed, lost. We had an outhouse. If you didn't make it back instantly to the house, you risked freezing to death, walking at night in the wrong direction. I remember we used to stuff newspapers under the old wooden doors. And even that was insufficient. See these spots on my hands? They never go away."

Taman let the image sink in, then asked, "What you said. It sounded Lithuanian?"

"Yes. On my maternal grandmother's side. Me and Simon."

Taman knew that they were fast approaching their first destination. He wanted somehow to prepare David Lev for stepping off the train. "He's been out there his whole life. It's all he knows, David."

"And you've read too many stories by Rudyard Kipling."

"Maybe, but if any of the pictures I've taken were to emerge in the worlds of social media, well, it would be disastrous, for all concerned."

Lev thought about that, now with new and striking evidence to bolster his utter bewilderment. Indeed, a wild child still out there. A Jew who had survived, who knew things... Scientifically, if anyone ever got their hands on it, the whole thing would surely backfire in ways entirely reminiscent of Nazi Germany. Corporations, military eager to explore new types of back-breeding, efforts at hybridization, eugenics. Genetically modified creatures had already been the rage for decades, with thousands of patents granted to new organisms.

And where did the Holocaust fit in – right in the middle, Lev realized.

And so did he.

***

At length, they had both grown silent. Lev had fallen asleep on the train, and had discovered in his heavy slumber the fact of that misery that is our kindred, our unsolicited partnership, in the name of the genome, with an unknown collective.

He seemed to dream of it, a thought so vague, distant, almost a murmur... That our species was forever plotting its gravestones within minutes of the consecration of our self-awareness; a gene in human evolution conditioned, and accelerating without our actually being aware of what was happening to ourselves from the inside – to be wary of the statistical minimum: the fact that no one was indispensible, unique, necessary. That our humility was not chosen, or accepted, but seminal. That we could not fly, nor run as fast, or survive with anywhere near the expertise and fluency of most other species. Even a housefly, with aerodynamic genius and myriad eyes.

That our legacy was wrought of fragility and broken promises. That our bond with other species, wild species, was never truly tested beyond those who accepted our company.

Chapter 55
Disaster

ERUPTION. The screeching of brakes, the moaning, the harrowing cries.

Lev was thrown forward – no seatbelts, of course – his head swishing the empty air, Taman falling back, then up towards the ceiling of the train, his head twisting in a furious angst of gymnastics.

Screams rocketed and ricocheted, just as the ticket police were again coming around, falling into a seat and cursing in every direction, then running toward the front of the train, which was trying, albeit kilometer by kilometer, to come to a halt.

All felt the crash. It was a big one.

Passing by them, outside the window, slowly now, a burst of flames, hay on fire, the dead cow, the dead man, the mangled body of one, possibly two, even three others in the cart.

"Jesus, Taman, there's blood everywhere," he muttered.

"We've hit a carriage. People are dead. See the cow! That's our unique black variegated, offspring of the Dutch Golshtin."

"We?" Lev uttered, his brain awash in the perplexed hyperbolics of yet another explosion in the company of this *techsupport*....

An explosion that, in a flash, seemed to target Lev's own guilt in this collusion of mechanics, the riders on the train, those who had, by sheer coincidence, conspired in the horror meted out so instantaneously in a universe not given to the least bit of fairness.

"A Dutch *what*?" Lev stammered.

"Golshtin. A cow breed. Jesus!" Taman hurled the words past his traveling companion, out toward the world they were entering, passing by, forced to meet head-on.

"What happened?" Lev repeated in a fiasco of slurred words, or perhaps had uttered just once. His shoulder had been crushed against the seat – fortunately the right shoulder, not his left, which had previously been injured in the café explosion.

...*I am so fucked*... Lev's brain dizzied, twirling like an out-of-control sparkler, his whole upper body reeling in pain.

In Russian, Polish, and Belarusian, people were calming one another. There had been no apparent injuries inside their cabin, a rarity considering they were in 2nd class. No children had been thrown.

"It can happen. But it is rare. They are wild, these Belarusian bovines," uttered Taman, still stuck on the cow. He loved cows.

"Hybrids, they stray, naturally. Perhaps homesick and searching for Holland, Denmark, where their genes originate. There are trains with feed. The animals come. Are you all right?"

"What are you talking about? What is this fixation with cows?" Lev had suffered a degree of whiplash. His joints were now burning, heavy with the throes of the train's impact, and he let out an expletive. It was all too much.

"We hit a horse and a cow and people."

Now there was shouting in many languages, and people were standing and moving about where slumber had all of two minutes before prevailed with the rigid equanimity of an uneventful train ride.

"The trains of Europe," Taman uttered. "They contain all the tragedies."

***

The calamitous event was over. An hour or more was spent examining and picking up the pieces. Authorities had combed the scene of the accident, and now, beyond redemption, the journey resumed. Nothing else could be done for the victims. Hundreds of busy passengers had destinations to reach. Life for most would go on.

An hour passed. Lev had peered down at the molten remains, seen the splayed, eviscerated full-bodied horror of two large Belarusian cows struck by a train at, what, 80 miles per hour? – not to mention the carnage that the poor bovines were trundling.

Apparently a draft horse was also involved. Incinerated in a microsecond.

Lev reclined, now with a crick in his neck, anxiety accelerating the entire picture of himself, as if framed during wartime, as much to forget as to contemplate. He tried diligently to doze off in a trail that bounced off any number of random events in his life.

He fell asleep directly into a stark ruin of a picture, the sum total of a syndrome, had the Holocaust kept going, that would have most certainly resulted in the total extinction of humankind. Atomic bombs would have been dropped everywhere. Cosmic Equilibrium Theories, but without a single instance of grace, of human nature at ease, rather than half-staff; a new human nature that worked together with nature, not to destroy her. Some Quixotic, final prayer.

Or not. Perhaps, as Mal had said, paraphrasing T.S. Eliot, a slow, agonizing whimper of an ending.

He could see the smoke. Hear the rain of molten lava. Germany and Kaiser Wilhelm II looming large, hiding out in Holland, a guest of Queen Beatrix, sending messages of encouragement to Hitler.

55 Disaster

The train took a perilous-feeling curve on the tracks at a speed that roused Lev.

He kept trying to force himself not to nod off. He was afraid of the sensation, one not unlike passing out. But he couldn't prevent going over that terrifying edge. He was worn down, exhausted, by everything.

Chapter 56
Scrimmage

"I need you and your wife in my office, if you have a moment."

"On our way," Jake instantly replied. His boss was never one to mince words.

Allan Hobbes greeted Jake and Claire, who came empty-handed, not even a pen between them.

"Have a seat. How are your kids?"

"They're doing well, thank you, Allan," Jake said nervously, abiding by the air of first names handed down in his department.

"So. You're watching a fast-moving train in your region, I presume?" Hobbes stated, casting a rough and tired glance down at the scattered mob of papers on his desk.

"We were," said Claire. "Until last night. A blizzard has closed in on the region. Our eyes are gone."

"Your reports have been erratic, to say the least."

Jake looked at his wife. This was a bombshell.

"So," Hobbes continued. "Your reports have been succinct but without substance. And that's understandable. But here's the thing: A lot of congestion down there. Seems to be a point of considerable focus, for Israeli, Russian, British, and, of course, Belarusian intelligence. But from even before the moment of that terrorist attack in Rio at the U.S. Consulate, we've grabbed dozens of communications from the four countries specific to that National Park, and – here's the really interesting part – Science and Technology over at State. They've maintained a stronger than normal interest in this. So it seems we have a problem."

"What kind of problem?" Claire asked.

"Genomes International, out of Minsk. Ever hear of them? I doubt it's the real name."

"No."

"I wouldn't have thought so. Here's the deal: They have somebody on the ground, we don't have a clue who, but the mole at State probably does. Is there any bloody new GIS technology we can hack into, anything, that would give us eyes at that research station?" Hobbes asked. "Jake, you have friends. I know you do, from all your days at NASA, your friend from Columbia?"

Jake relayed the query to Claire, like a relay race. She, too, rather dodged the inimitability of this overseer. "The weather is slated to continue like this for two weeks, unprecedented snowfall. We can't see a thing."

"We're locked out," Jake attested. From space, one could imagine this huge portion of Eastern Europe under a level-five hurricane.

"What about heat signatures?"

"The storm layer is two miles high. Yes, there is thermally detectible movement, but it's a National Park. A lot of cows."

"Bison," Claire interjected. "Wisent, specifically. And other terrestrial mammals, obviously."

"Well, eyes and ears to the ground," Hobbes concluded.

As Jake and Claire were leaving his office, Hobbes added, "Oh, thanks for turning down that invite. I expected nothing less, of course. But I have to tell you, you both missed a hell of a good lunch."

Chapter 57
The Great-aunt

"My great-aunt knew him as a child."

"What are you talking about? I don't believe you."

"It's the truth."

The train had slowed to half its normal speed after the accident. They were about an hour away from their destination.

"Who is your great-aunt?"

"She died over two years ago."

"What was her name?"

"Sarah."

"Sarah what?" Lev asked intently.

"My last name. As for her maiden name, I don't actually remember."

"How old was she?"

"A few years younger than yourself."

"From which country?"

"Poland."

"What town?"

"Your town. It doesn't exist anymore."

"What else?"

"She explained some things before she passed away, before you and I met. Most importantly, it turns out – and I hadn't ever seen it while she was alive – she kept a diary, which I have hidden. It names names. I was lucky to locate it, before others nabbed it."

"Names?"

"It's likely that he saw everyone who was involved in the massacre. They are all buried there. This is something huge. International. There is no backing away from such a truth. My family lost people as well in the war. At least one is also anonymously buried out there, I believe."

"You are too young to know anything about it. I came from there. I was born there. I saw everything, up until, more or less, a couple of years before the end. But I was the lucky one. I got out. Pure chance, nothing more. An uncle in academia from Dublin. Utter, random good luck, for me and my uncle. Not for the rest of my family."

"I wasn't born yet. That's true. But never too young. I still live here. Surrounded by it in so many ways. These echoes surface, as you know. From France to Hungary to the old baker catty-corner from my duplex, a relative, who lost members of his own family. Whereas you, in sunny California, I can only imagine."

"It's not what it's cracked up to be. Anyway, how could you possibly have put all this together?"

"I've gone to every Jewish Web site. One puts two and two together. The diary makes it all incontrovertible. And then, the bunny rabbit."

"Bunny rabbit?"

"I thought, initially, that rabbits, or two species of squirrel, small hypocarnivores, were devouring the specific fungi in my experimental plots. Because there are no other footprints other than his. Wild boar, obviously the wisent; they would have destroyed the plots."

"Your monitoring sites have tape, numbers posted?"

"No. Impossible to distinguish from surrounding deep forest. When I saw him eating, I shot the photograph. Then I realized who he was, afterwards."

"And you made contact with him?"

"He saw me. Ran – "scampered" is the better term. He is hiding out, David. And my great-aunt's diary makes that scenario a plausible one. More than plausible."

"Where? He would..." Lev's mind was racing, trying to calculate the incalculable, "he would be eighty-two years old now. Older, even. My only living relation, besides my wife. Will we find him?"

Chapter 58
Lev's Dream

Miraculously, and to the perdurable bad luck of all others, we managed to muddle through World War II. The Jews continued, in every respect, despite their unspeakable tragedy.

These thoughts were coming to him, through him, without any sense of the real world. Everything he believed, held dear, studied, outside, inside the realm of his own awakenings, all collided in a strange yet singular commentary that made him grind his teeth, pulling at the jawbones, however many there are...

Fragments of some Peaceable Kingdom floated across his faintly-breathing sleep, that vast imaginary litany of the Prophet Isaiah... of the Velvet Breughel's many Paradise scenes... an annual Westminster Kennel Club, and a motley hallucinogenic figure of the man Roosevelt, having concluded two presidencies, who was paid in today's dollars $1.3 million to report on his famed 1909 Smithsonian Expedition to Africa in which with colleagues thousands of specimens were bagged, including over 500 large mammals... the sheer... the sheer *what* of it?

The man, Roosevelt... clouded in a confusion of so many contradictions that Lev had managed to study over the years. Saving what would eventually amount to nearly 600 million acres, a hunter doing that.

And thirty years later, another hunter, insane, psychotic, also with vast, vested power over an entire nation, slaughtering six million Jews. The totally blurred set of boundaries within whatever notion of sanity may be laid claim to, validated, contested by human nature.

He thought he heard his own words, from some lecture – no doubt laden with cuss words – somewhere, sometime back, that any fool can slaughter another. Slaughter doth not delineate, his head vibrated, in and out of the nervous apprehension that is our collective turbulence; perturbations pedaling atop limbo; sacrifice and terror trembling upon the chasm of a conscious cancer.

There before him were the wisent, approaching ever so gingerly. The huge forest bison of Bialowieza...

Lev's mind went back to those days, with the added quanta of what he'd learned about the fabulist creatures, creatures of ancient lore and imagination.

The buffalo was tied to a tribe, not unlike his own people: the Bovini; Family, Bovidae. Aristotle and Linnaeus had chimed in on these creatures, just as the American Indians, Canadian First Nations Peoples, African nomads, Tibetan yakherders, the vegetarian Todas of South India worshipping their rare river buffalo, and cave painters of the Paleolithic-Upper Aurignacian, had all paid tribute to this global tribe.

These vastly sentient beings that David vividly recalled from his few encounters during childhood, meant more than all the confusion between the biologically distinct genera of bison and buffalo, antelope and cows could ever say. They went to the heart of Bialowieza. Without the wisent, there was no forest. Without the forest, no wisent.

***

Lev's left eye popped open, but only in a blinking alpha state, and what he saw meant little to him. A man sitting across the aisle. He did not know him, of course. A meaningless eye contact.

A landscape out the fleeting windows, the windows themselves, that he wished not to recall, as if they somehow intimated a bitter antediluvian deluge.

Then re-immersed. The throb of the continuing metallic train noise and occasional high whistle. A dream he was not able to give up . . .

He sat up and started to cough, covering his mouth.

He glanced at the face of his watch.

"You all right?" asked Taman. "You were sleeping. That's good. You're a tired man. And it's just beginning."

"It's obviously quite cold out there."

"Yes." Lev touched the ice forming on the inside of the window, crystalline, and seemingly pure.

Taman sat motionless, fixing his gaze out the window at the passing world. He worried for this American, for his own family. For everyone.

And for Lev there was nowhere to take these melancholic musings. He was left with the reality of a defunct species that abided within himself. An aboriginal dreaming, of sorts.

All this reflection, ruminations born of exhaustion, had landed the professor directly on the burning image of his little brother. It was a powerful ache that had opened his eyes.

And an image etched into copper by the artist Dürer, which he knew well, a concept that had never left his heart: the so-called "Eustachium" of the medieval hunter Saint Eustace, on his horse, seeing a vision of a great stag with massive-sized antlers – a Renaissance version of some Boone and Crockett Club trophy, to which a Theodore Roosevelt had lent heavy modern compulsion – and then, suddenly, behind the deer, a glowing little Christ being crucified . . . or something like that.

A vision profound enough to change the hunter into what could only be described as a humble St. Francis type.

No more killing. If only Christ's message had impacted others. If only such mythography would have us believe in "The Vision of St. Eustace," which was probably first painted, Lev knew, didn't everybody know? by Pisanello in the mid-fifteenth century. It hung in the National Gallery in London, where he had often been dragged by Sasha in later years – it was one of her favorites, as well – after the ritual of feeding the pigeons down below, in Trafalgar Square.

If only that painting had hung everywhere, in everybody's heart. But it did not.

He even felt foolish reflecting on these hugely distant, nearly irrelevant aesthetics, human nature being essentially fucked, in his opinion; not Christ-like.

Eventually, Lev again dozed off. At his age he knew these brief respites were precious. And almost immediately, Taman realized, the old man was snoring. That was a gift, thought Taman, who had great difficulty sleeping, always.

... There were he and his younger brother, Simon. They had, in fact, climbed a tree together outside their village, the village that no longer had a name. Memories put out like a cigarette upon a gutter. A tree under which they had frequently played, a giant Norway Maple.

... The anxious late night and eyes heavily half-lidded with serious psychosis – the history of Europe in the twentieth century, its blasphemous day by day plastered in so many scenes of atrocity, beheadings, crucifixions, lamentations, stoic Madonnas, wintry scenes laden with oppressed figures, battened down crows on trees forever bare, bayonets and AK-47's, in every art museum throughout Europe's great cities – was suddenly there before his mind's eye: the castaways of memory, the blur of blind schizophrenia receding like wave upon bashing wave of guilt, and forlorn such as only the staying power of a tired ocean in the mind of a fatiguing old soul can conjure home again. A Chopin-like nocturne.

... The never-never world of the Ba'al Shem Tov, of Hasidic loveliness that was his fantastical mud-splattered decrepit wooden and tumbling brick village, homey, humbled by antagonism on every side of geography and history, but home, and – for all that was and is no more in the far-off childhood of David Lev – had long ago evaporated, only now to begin the journey of return in an emerging sleet, cold and miserable, of reminiscence right before him.

© M.C. Tobias

Chapter 59
225063 Kamieniuki. Region: Brest

"We're only having dinner, not checking in," Taman advised the professor with a knowing look, a repeat performance.

He then explained how everyone knew everyone, but the local police were from the region, and Taman was not. Moreover, they knew, and were undoubtedly envious of the fact that he would earn more than they could ever hope to earn, although the quadratic differential – Lev realized that Taman had an annoying habit by now of characterizing basic things in mathematical complexities – by American standards, was trivial. Were they to check in at the hotel, it would instantly trigger suspicions as it was known by police, those in the hotel, at the nearby market, post office, and bank that Chernichevsky never stayed at the hotel. He had his own lodgings, a research station at the edge of the National Park. But dinner was *de rigueur*, a show of support by the Academy of Sciences for the local economy.

The wind was howling, a persistent winter, with far deeper consequences here in the southwestern portion of Belarus than where they had originated in Minsk. Throughout their train journey the weather had progressively worsened.

"Don't say a word that gives away the fact that you're American," Taman informed Lev. "You actually should not be here, an American, you understand? It's a military zone, in actual fact, this time of year."

"French good enough?"

"Yes. French is acceptable. Diplomatic relations are easier."

Lev nodded, as they sat down at Taman's customary seating in the rear.

The waiter, a long-time acquaintance of Taman, although not what one would call "a friend," was quick to attend to them. He had absolutely nothing better to do. It was not a busy night. Indeed, Taman and Lev were the only two in the hotel restaurant.

"*Piotr, jak ty? Dobra, Tamań. Viarnucca tak chutka. Tak. Usio dobra? Dakladna. Tak što rychtavać sionnia viečaram?*" (Peter, how are you? Fine, Taman. Back so soon?)

"Yeah. Everything good? For sure. So what's cooking tonight? I think we can easily have your regular in 15 minutes. As you can see, it's that time of the year. You're probably our only guests this evening. Vegetable *draniki*? Yes. *Bulbashi*?

Of course. *Belaya vezha* salad, just the mushrooms and onions, no cucumbers or eggs, lots of sour cream and cabbage. And give us some fried potatoes – extra burnt – my friend here, from Paris, has never had food from Belarus, he'll love that. And your buckwheat kasha. And to drink?"

"You are French?"

"Oui. Et vous?"

"What would you like to drink, professeur?" Taman injected, before familiarities reached a critical mass.

"Bordeaux?"

The waiter immediately suggested either the Knyazheskoye or Merula, both excellent natural red grapes, quite popular.

"There are something like 135 fine grape varieties in Belarus," Taman chimed in.

"Bravo. Pas de Bordeaux?" Lev added.

"Merula," the waiter repeated.

"Yes, perfect," Taman said, then excused himself for a moment. He had some arrangements to ensure and went to see someone in the hotel, someone he could trust.

Lev was brought a glass of wine, poured there in front of him in the traditional sampler.

"Very nice. Merci!" This time, the stupid French intonation. Mundane words emerging like bean sprouts, senseless in the horrid scheme of things. And dangerous, lest the waiter start speaking French. Lev could read it. His speaking, however, was disastrous.

Lev sat there alone and considered the prospect looming immediately before him. The very food, names among which *draniki* and *bulbashi* were familiar from the queasy proximity of his upbringing, but words he rarely heard in the intervening 75 years, except at places like Nate 'n Al's in Beverly Hills.

As the waiter departed for the kitchen, Lev felt himself slipping into some other place, some other time. He rubbed his eyes.

...Here am I, on the edge of a vast primeval forest. I am myself an old man. With perpetual weariness. And I still remember the images of Adolf Eichmann, having been grabbed by Israeli agents in Argentina, in his glass cell at the trial in Jerusalem. I wish only to fall asleep and make it all go away, these impossible contradictions. The red wine, no less than three glassfulls. A fine warm dinner...

Disorienting jolts, the sadly chosen ones, hammering home the fact that, indeed, he was home.

A dinner that would erase the stinging perpetual pitfalls of memory that have always been deemed crucial by the world's oppressed... we shall never forget. For Jews, even Jews of odd assimilation, like Lev, that included place names he knew from years across Western Europe, with binoculars, in meetings, but also out of a sense of necessary moral homage to the compass of cruelty about which his family's erased destiny had been so intertwined.

Indeed, from what he had lived with, until now, he was the very last of them, a sensation, perhaps, that had nurtured his very profession, the study of extinctions. Personal histories.

Chapter 60
The Histories

Lev had been taken by his uncle to the Ballybough Jewish Cemetery in Dublin long ago, a place at that time behind some two-story building at 67 Fairview Strand Road, but said to be the third-oldest Jewish cemetery in all of Ireland and the British Isles. He had visited the Ashkenazi Cemetery in Paris, first opened in 1785; wandered the Judengasse in Austrian Eisenstadt where, he had been told by a taxi driver, "no Jews were allowed" – and that was in the late 1960s.

"Allowed?" Lev had questioned the driver, with an obvious attitude. Backing away, slightly, the driver, who spoke fluent English, tried to clarify, probably realizing he had obviously picked up a Jewish customer: "You see, no fault of the locals, but no Jews returned after the war."

But, in fact, Lev had learned, for over 300 years Jews had lived and flourished in over thirty households there, beneath the castle of Prince Esterházy. Indeed, Austrian Jews under Franz Josef I had probably had it better than most Jews, ever, in their history, as described once by Stefan Zweig. They purchased the daily newspaper in Vienna, unconcerned with worldly affairs, focusing instead on which operas were being performed that evening. What a life. A fleeting one, at best; and a rare historical exception to the rule.

And the Emperor, dying peacefully at Schönbrunn, his 1,441-room baroque palace. But for every palace, there were hundreds of thousands of gravestones, usually leaning, twisted, vandalized. From Belgium to Italy, Lev had made pilgrimages to those countless yards of lost souls and cumulative heartache. Whole communities, like the one he originated in, that had ceased to exist for all the wrong reasons in human affairs.

He had been in Trent on the very day, in late October, 1965 – he remembered for an obvious reason – that the Catholic Church recanted its campaign against the killing of Saint Simon by twelve Jews, the so-called "Jewish ritual murder" in 1475. Twelve Jews had been falsely accused and executed.

When the recant order was published, the exhibition of a mummified, approximately 30-month-old St. Simon, stained glass windows depicting the alleged martyrdom of the saint, other relics, and the very chapel devoted to Simon, were all closed down and replaced by windows representing Moses. There was a ban on the very mention of Saint Simon during mass throughout the city, the sacred liturgy that stated "*Auferte gentem perfidam credentium de finibus*" (Take the faithless tribe from the borders of the believing).

A similar heresy alleged that Saint Christopher was killed in a Jewish ritual murder, the flagrant rumors of which infected Spaniards and resulted in the edict of Ferdinand and Isabella, and the Inquisition and expulsion of Jews from Spain.

Same old heinous fabrications of convenience lurking around virtually every European capital, village, and history book.

But it was during Lev's trip taking him to a region within the outer pale of his own birthplace, to a conference in Budapest, that he now reflected upon the most salient intersections of culture, ecology, politics, and Judaism, smacking of the very moment.

To Buda, more precisely. The synagogue, which the Hungarian official tourism map referred to as a "Chapel" on Castle Hill. Number 54 came into his and Sasha's grasp, there among the most chic and elegant of homes. One 6,000-square-foot, two-floor apartment overlooking the Danube, selling for nearly thirty million euros. They had tried and failed to see the remnants, beyond the stones at Kozepkori Zsido Imahaz, at 26 Tàncsics Mihàly utca (street).

But the crucial links to a medieval Jewish truth – a synagogue that had evidently been built as early as 1366 and destroyed during the late seventeenth century during Hungary's war with the Turks – were encinctured within the confines of enigmatically private property. Several apartments. No harm. Life goes on. But the museum itself – enshrining the old synagogue that had only been discovered three centuries after its construction – was closed "temporarily."

He and Sasha could not consummate their quest, or were just stupid about the procedure. They walked past the singular tree separating the "chapel" from the large Lutheran Church. Across the triangular park stood the Hungarian National Archives, a monolithic but attractive enough building.

Hungary was known for having had the largest number of Jews in proportion to the total population of the country, slaughtered during the Holocaust – well over 550,000 murdered people.

Half a mile down the street, southward, the changing of the guards at the Presidential Palace. Goose steps. Rifles. Pomp. The swirl of so many architects, wars and revolutions, grandiose expenditures, and defiant egos. It was there, in one of many deceptively inviting cafes that project an air of Paris along the length of Buda, in any weather, that Lev had one day ceased his compulsive dalliance with birds in order to have several Irish coffees and read two novellas by one of Hungary's Nobel Laureates in literature, Imre Kertész, and sense in both *Liquidation* and *Kaddish for an Unborn Child*, books that he raced through without actually wanting to read every word, just obtain a gist, he had been struck, as by some enviable lance, if such a thing can be said to happen – it did, it does, it must, to those who are blessed and cursed – by the line cited by the esteemed author of Calderon's *Life Is a Dream*, to the effect that man's first offense, or crime, was the very fact of having been born.

And, sated by these unspeakably complex self-annihilations of reflection, to then take in, as Lev remembered doing, while Sasha slept in at the hotel, recovering from jetlag, which she suffered in greater bouts than her husband, the incessant sense of depression marking that insensate embellishment of history's false bastions so elegantly enshrined atop Buda.

As if to emphasize the characteristic fool's paradise that was the Dance of Death in art, in nature, and in all human affairs.

An invasion of pathos, now in grim retrospect. But at the time, after four Irish coffees in succession, and a sampler tray of dumplings doused in plumb sauce, of little more than a romantically passing interest in the perpetual drizzle.

It was the same Hungary in which, on the one hand, at the Hungarian Academy of Sciences he had seen the fourteenth-century illustrated *Hagaddah*, donated by the mother-in-law of David Kaufmann, the nineteenth-century professor in Budapest at the Jewish Theological Seminary. A city in which Theodor Herzl had been born; a city commemorating Raoul Wallenberg's rescue of over 600 Jews during the Holocaust when Budapest's largest synagogue, at Dohány Utca #2, which had seating for a congregation of 3,000, was turned into a veritable ghetto by the Nazis.

Elsewhere, at the Jewish Museum, over 2,000 Jewish martyrs lay in a mass grave.

What Lev remembered most, however, was the massive sculpture of that unlikely savior of the Huns, and hence – of Hungary – the mythical, eagle-like Turul. A raptor that had been woven for well over a millennium into the migratory tribal legends describing those peoples that ultimately settled in Hungarian lands; a bird of the same size, Lev recognized, as the extinct New Zealand Haast's eagle, the largest flying avian since the time of the Pterodactyl.

Today, in Hungary, the Turul had become a key symbol of the right-wing political platforms. If the economic tide worsened, a quietly anxious woman selling books told him, there might be nothing to stop another Hitler from arousing the masses.

All seen through the much warmer than usual rain of an early December. Warmest on record. And that was the last time Lev had ventured anywhere near his homeland, until now.

***

Taman returned to the dinner table, where the two men sat by themselves and quietly ate a delicious meal, every bite bringing home more and more residual memories for Lev. Tastes that were like relic artifacts of Lev's forgotten narrative, the very bone marrow of his grandparents; food so entirely different from that in what he would soon come to call "home" in Ireland.

An odyssey rich with discord, particularly upon entrance to the New World that was America, in the sense of having been systematically disengaged. The weight of that lifetime of unconnectedness, of having forgotten most of whatever Yiddish and Hebrew, Lithuanian and Russian he may once have known, save for some of the linguistic stanchions of his faith.

While he never had a bar mitzvah, and was thus no "son of the commandment," Lev had been confirmed, and could never forget that pillar of a prayer, "Shema Yisrael...Hear, [O] Israel: The Lord our God, the Lord is one," "אֶחָד 'ה אֱלֹהֵינוּ 'ה יִשְׂרָאֵל מַעֲשׂ," Deuteronomy 6:4-9, the first carefully documented declaration of monotheism in human history.

Aside from those basics, Lev was one more fully assimilated outcast, like millions of the diluted minds and deeply distraught hearts of the Diaspora.

He tried to shake off these pangs.

"How far, Taman? How far are we from that tree, just now?"

"About thirty-five miles. We'll be hiking, of course. It's not easy. You are up to it, I know that. I have two ski poles for you at the research station. They will help. Have you ever used snowshoes?"

"Long ago. Very cumbersome."

"We have them. And gaiters."

"We'll be leaving a lot of tracks," Lev pointed out.

"They are predicting heavy snow. We will possibly leave no tracks. And your shoulder?"

"I'll manage, somehow."

"Will we be falsely checking into this hotel, like the last one?"

"There is no point. If we are followed into the forest, well, we must talk about that. But not here."

Taman paid the bill, complimented the chef, said good-bye – along with Lev's au revoirs – embarrassment bringing a first near-grin to Taman's otherwise flat-affect of a face; and suddenly the rightly perplexed waiter (*too much Merula for a mere Frenchmen*, he probably thought), and they walked out the front door of the hotel.

Within minutes a young woman driving a banged-up and muddied Lada Priora arrived to pick them up.

All three in the car, Taman introduced the professor to Ulyana, not a graduate student but a kind of manager whose job it was to look after their research station at the edge of the forest all year round. Her boyfriend Jakob would come to visit periodically, and she earned spare income from leading the occasional tourists into the National Park, not this time of year, of course.

"Ulyana has been here for over two years," Taman said, as they drove the twenty-two kilometers in the dark, from tarmac to a mud road covered in heavier and heavier slush. She spoke no English.

A deer sprinted across the headlights, then a second and a third one, ghostly white. Lev could see how deep the snow to either side was. They had arrived at the Belovezhskaya Pushcha National Park. A sign was distinctly visible. A few buildings emerged in their headlights as it began to snow lightly, presaging a long, banging night of a raging blizzard.

Indeed, Lev saw all the instantaneous signs: the wind was picking up, and within minutes the snowflakes seemed to change into the size of enormous cotton balls like those of the recently fruited Chinese silk floss trees that graced portions of Los Angeles. There was one on Lev and Sasha's street spewing its warm downy nesting material for robins and magpies alike just days before Lev had left for London.

For Professor David Lev, the snow coming down mirrored an evolutionary seizure affecting his eyesight, willpower, point of reference. He was dizzy because it presented so intimate a realm of recollections from childhood.

In the back of his mind, a chain of events, biological corollaries, an all-consuming trepidation for the man who had originated in these very climes. The thought of venturing out into that darkness, beyond the confines of a modern automobile.... *To go on foot*... The notion that – out there, somewhere, was his brother. He could not stomach any of it.

No being rushes into the fire, wishes to be accused, stung, overwhelmed. Time was punishing enough. But this was an activist's madness, the eternal return academicians were so fond of glibly citing, without the slightest clue as to what it actually could mean.

Anthropologists in the Amazon listening to a macaw to grasp the lost language of an extinct tribe, archaeologists in Israel uncovering tunnels from the time of King David, physicists at CERN – the European Organization for Nuclear Research – endeavoring to locate a single, all-encompassing particle that could explain black matter and the origins of the universe. Like the unstoppable Higgs boson.

But for Lev, such unequivocal and absolute closures on the reigning questions of our time, of human nature, were windows on the monumental tipping point of sadness. His own poverty of spirit that conspired at this very moment to sap his will. He was terrified.

Are we, was he not, after all, his brother's keeper? More to the point, how could a Jewish kid – the child that was Simon, born into the final years of the *shtetl* world and raised during the precise moment that the worst cultural vendetta, a political sledge hammer that was ever demonically devised – how could such a boy become a man up in the air, living in a tree among bark beetles, woodpeckers, wolves, and bison – the unique and critically endangered Bialowieza bison, or wisent?

How could a Jew survive in the guise of a wild child? That was the fundamental flaw in every second that had passed since this insane techsupport first invaded David Lev's life.

And if he were to answer in the affirmative his life-affirming commitment to Simon, then what? What were the options? How was the trauma to be dealt with, on all sides, by everyone who must necessarily get involved? Or, conversely, let go of the very temptation to get any more immersed?

These pangs of doubt no longer presented an option. He was on the geographical cusp of D-day.

Chapter 61
The Watch

Ivan Petrovsky had been waiting. Waiting a long time for this.

He'd seen Ulyana leave for town in the dark from a craftily outfitted bunker of sod, broken fir branches deep in snow, and a superbly fashioned bivouac of ferns and mosses. Downright toasty, with the perfect vantage point over the rear entrance, or exit, depending on one's perspective, of the entire research station.

On one occasion, a wild boar had tried to dig it out, although unsuccessfully. No truffles, only a blast of aerosol spray to the face containing a mixture of red pepper and a harmless solvent. Boars hate pepper.

And he now finally had discovered Ulyana's caches. Not one, not two, but three high-powered rifles: a 223 Rem – a favorite with NATO forces – that she kept beneath a pile of used batteries and recyclables, never disturbed in the winter; an SR-25 7.62, also known in the parlance as the MK-11 that she'd placed craftily in the rafters of a nearby wood-mounted outhouse, and the US Navy 5.56. That one he could not tell where she kept, except on her person, and probably had it in the vehicle that night.

He'd twice seen her practicing in the forest with what he reckoned must be some kind of AAC Cyclone sound suppressor. She did so at great risk.

One day, Taman – heading back to the research station from one of his monitoring plots – nearly missed chancing upon her. There was no telling what might have happened in that instance. She hadn't simply been turned. She'd been groomed, probably even before entering graduate school. Her whole masters in dendrology was a front, which would explain how she obtained her degree in less than a year, by his clock; and why she spent one day a week also leading bicycle eco-tourists in season across the border area, where Petrovsky also earned his post-doc, and some spare change.

She was playing a complicated game. There was only one way in hell that she could ever have obtained the weapons, or the training to properly use them. And there was no doubt that Petrovsky's mother's previous U.S. $3,000 cash disbursements were going to her. Petrovsky had managed to follow her into Kemieniuki one day, where she withdrew one hell of a lot of cash from one of the few functioning ATM machines in town.

He could not spot the precise amount, despite a fine pair of birding binoculars, for fear of being recognized. People knew him in town after three years at the research station. But for more than eight months he had cultivated a careful anonymity. The timing of Ulyana's cash withdrawal – the very demeanor of her withdrawal – spoke volumes. She had to have had access to sums beyond the normal one-time amount, which was 1,500 BYR, or U.S. $100.

He had no doubt whatsoever. She was working for John Vespers, which meant that local authorities had somehow been informed: she was inviolable.

It made a global solution far easier for Ivan Petrovsky to choreograph, as he dug in for another night, watching.

Chapter 62
The HIT

It was a brilliant late spring afternoon along the Potomac at Riverside Park. Jake had jogged four or so miles along the Bootlegger's Trail and was walking back along the famed Madison Escape Trail, while Claire and Jake's parents were packing up a leisurely picnic with the twins amid blooming Virginia Bluebells. Red-headed ducks delighted the little girls. And there was an osprey who was impressively fishing from a comfy spot atop a wooden post out on the water's edge.

"You better call your husband. He has never once been on time. Can't you break that habit?" Jake's mother went on.

Claire picked up her cell and called Jake. "It's nearly four and your mom's pissed."

"Eddie, don't go away, it's my wife, just, just one sec – sweetheart, I'll be there in a minute. I can see you. I'm a thousand feet away. Tell my mom."

"Eddie, come again?" Jake said, speaking in the late day filigree of quiet solace at the base of a large chestnut oak tree.

"I have to be brief. Just got wind of it last week. A virtual organization, 'HIT' for short."

"Did you say 'hit'?"

"Yes. I'd never heard of it, can't find it online. But they exist, alright. The acronym stands for the Halacha International Trust."

"What is it?"

"*Halacha* is Jewish law. So your guy in the forest did thorough homework. He's not just a mushroom geek. Apparently he tapped into numerous Holocaust survivor data files. And he found one. Sent a couple of images to some famous American scientist."

"Of what? To whom?"

"A university research library system. Heavily encoded."

"What's in the pictures?"

"Don't know. They were vanishing images."

"What, like snapchat-ish, five seconds?"

"No, two minutes. But still fast enough to be untraceable. Anyway, lad, all connected, royally fucked up, and much bigger than either you or me. Be careful. Bye now."

"Eddie, wait, are you guys … ?" but Eddie had already hung up.

Chapter 63
Convergences

By the time Jake and Eddie were received, it was not into Allan Hobbes' office, but the Agency's Deputy Director of Joint Operations' special meeting room. Basically a bunker. Cosgrove and Cosgrove Upton had only met the Deputy Director once before, the day each was hired, and that was in the cafeteria. Claire had had no problem that day devouring a veggieburger, loads of mustard, and a tomato whose seeds had squirted into the DD's face, in "an otherwise uneventful meeting," Jake delighted in recounting that day.

"She never saw it coming, sweetheart. You nailed her! Paintball is in your future."

On this day there was a total of four people in a room that could hold thirty, five stories below ground level, past all the data crunching machines, glassed in and fully inaudible to the outside world. Jake and Claire, Hobbes, and the DD, Dr. Helen Norwich.

"This will be brief. We have direct knowledge that there is a shitstorm unfolding beneath what Jake, you, and Claire know to be the worst storm to hit southwestern Belarus in a century. Allan has walked me through the many incongruencies. The State Department is in a peculiar situation. Their commercial arm has entered into a rather ambitious, if nebulous, coalition concerning a revenue-sharing deal for some new allegedly outrageous genetic discovery. But not a single individual at White House Science and Technology, DEA, NIH, or our friends over at US Fish & Wildlife can verify it."

Hobbes added, "That would include our sources at the National Center for Biotech Info, PubMed, OMIM, GenBank, etc."

"You'll excuse me," Jake started, "but what is this all about?"

"What are we doing here?" Claire said, to point.

"You've seen several murders, or assassinations in that forest," the DD stated calmly. "What you've seen apparently segues from a terrorist attack at a café in Rio, followed by two more attacks, one en route to, and one within the U.S. Consulate in Rio. In all three of those attacks, the intended victom, a rather famous ecologist from southern California was involved and we believe is in the middle of all this. We didn't know why until last night."

The DD turned her glance to Hobbes, who in turn cast his penetrating laser into Jake's eyes.

"You see, Jake, we've always recorded your private phone conversations. Yes, even in your little girls' room."

"Fuck that!" Claire exclaimed. "How dare you!"

"Dear," Norwich began. "It's national security. It's what we have to do. Read your contract with the Agency."

"That's good. Then you know about Eddie."

"Jake, I've known about Eddie for years."

"So what you heard last night, his call to me, what do you make of it?"

Hobbes looked to Norwich. Claire was silently fuming.

"The problem is a certain David Lev, professor emeritus at UCLA. He's there now with the Belarusian scientist who was in that terrorist attack with Lev in a bar during the U.N. Summit in Rio three years ago. Somebody has been trying to kill both these men. They are not collaborating in any way. Different fields entirely."

"So what's their connection?"

"Jake, Dr. Lev is getting towards ninety. He apparently lost his entire family in the Holocaust, all but one, we think his younger brother. Since 1979 the U.S. Justice Department has been hunting Nazis in the U.S. who might be receiving Social Security benefits. Tens of millions of dollars were unwittingly given out to these escaped criminals. We can assume there are dozens of Nazis still freely circulating in Belarus and Poland. We followed the HIT, what Eddie was telling you about yesterday. You understand?"

"Right."

Norwich clarified, "There are any number of scenarios that could play out down there in that storm. Somebody is going to try to kill Dr. Lev to keep him from getting to his brother. We think that's the purpose of his visit, certainly not bird-watching, which is what his visa application reads."

"I'd think they'd go for the brother." Claire hastened to deduce. "I mean, if he's alive, he can identify Nazi collaborators."

"Yes, but they wouldn't want him dead. They'd want him totally hidden and unconscious."

"I don't understand," Jake stated flatly.

"Apparently, the consensus is that he's been their dream trial," Norwich continued.

"She doesn't mean legal trial," Hobbes added. "And therein lies the rub."

Norwich went on: "The Belarusian scientist discovered something out there, something evidently important to drug therapy. Off the charts important. Like all new drugs, there are the laborious and expensive phases, six clinical trials, more or less. Lev's brother appears to embody all six trials, over most of a lifetime, in one person. I've personally spoken with the Under Secretary over at the State Department for Economic Growth, which as you know supports environment-related issues. They've been monitoring all U.S. Consular affairs in Minsk that entail the slightest dangling modifiers. *Yes*, there are several multinational deals that have been struck, under, I would argue, the misleading guise of public/private partnerships. *No*, they

63 Convergences

claim to know nothing about a Holocaust survivor in that National Park – my connection strenuously pointed out that that strayed far from their portfolio, nor will they discuss anything that concerns Nazi-hunting. Not their bailiwick."

"They wouldn't anyway," said Hobbes. "State's got its own intel-gathering machine, with a budget five times that of the Agency."

"More importantly, I think, they are, strictly speaking, conflicted," said the DD. "I don't think they even quite realize how problematic a situation they're in. It pits Interpol and the Justice Department against the U.S. Department of State, and our Agency, and the FBI against the NIH. Not to mention the extreme need-to-know of dozens of international NGOs, like this HIT Eddie was talking about, on his way home in the subway from the Doughnut, by the way – in case you were wondering about the noise, Jake – he was near Paddington Station."

"Great. Good to know," Jake said with distinctly frustrated humility.

"So I ask again," interjected Claire. "What is this meeting about, other than the revelation kindly shared with me and my husband which, I'm sure" – she looked wild-eyed at Jake – "is going to preclude all future semblance of privacy or any family life on our part?"

"That would probably include sex life and pot," Jake threw in.

"Jake, Claire, it was temporary. No, we don't spy on our agents, as a rule. But I knew about Eddie, and, quite frankly, had some concerns about your own personal safety, the two of you. Your kids. Even your parents. As you know, we've not implemented any security procedures, as yet, in that domain."

"Great intel. Thanks a lot." Claire felt like turning in her badge right then and there.

"So what can we do?" Jake asked. He reached over and put a hand on his wife's hand to calm her down.

Norwich looked to Hobbes, who volunteered, "We have this new app, just got it in; the Agency hasn't tested it yet."

"What kind of app?" Claire pressed.

"It was developed in the private sector, never mind by whom; well, it's basically like the finest lens on the Hubble and can see through anything. We finally have eyes on the ground. And ears."

"What? How?"

"Down to less than an angstrom, ultimately – starting from half-inch increments, in bright light or in darkness; through a blizzard, a hurricane, a tidal wave, you name it," Hobbes said, with a near smile. "From 400 miles in space to the bottom of the ocean."

"That's impossible," Jake muttered.

"They've apparently aggregated every known algorithm in every satellite, used a multiplier effect much like filling in a field or frame for added pixal strength, and ..."

"And you mentioned audio?" Claire broke in. "Even among sci-fi aficionados, that would be stretching it."

"It *is* experimental. There could easily be glitches, although we are told that it is now fully operational. And, under the circumstances, we thought you two might like to be the first to try it out here at the Agency," the DD concluded.

Chapter 64
The Coming Storm

A figure hidden deep in snow saw the lights of the vehicle arriving and heard three voices. Two he knew well; the third he hadn't a clue about but with night vision binoculars he saw at once it was a very old man, he had no idea who.

Once the trio went indoors, he kept watching, waiting, certain there would be others trailing behind, sometime during the night, or the next day, or the following night. He was well provisioned, including a stockpile of some generic drug purchased from a pharmacist friend sadly going out of business in Minsk months before. His friend had sworn the drug had all the equivalent boosts to the system of modafinil, but – in liquid form, injected into a vein – had an even longer lasting impact than that used by the American military.

***

Lev had not slept in many nights. He tried hard to stop thinking about anything, but that had proved impossible – a retreat, willed, unwilled, into a nap resulted in calamitous memories colliding with superconductivity he could not stop; his awareness was a refuge bin of fears and old dramas he had tried so hard to outlive – as they reached the research station, emptied their few things, and went indoors, Lev being shown his single sagging mattress close to the floor, the location of a toilet, the fridge and coffee pot. Dishes on the open oak shelving above. Mice droppings generously distributed. And ossified bugs from the last days of the previous fall populating corners of every wall and cupboard.

A dead preying mantis, frozen raptorial legs, labrum and compound-eyes wide open, like a still threatening fossil, covered in frost that no human interior could keep out.

There was a detached annex that served as the administration building. Adjoining it, a dimly lit cold concrete structure in which there were two high-powered microscopes, the somewhat dated three-temperature thermal cycler for PCR (polymerase chain reaction) by which DNA samples could be amplified, storage units for specimens – thousands of them: lichens, mushrooms, tubers, a plethora of strange, dried, colorful forest flora arrayed in a variety of housings, on shelves, atop large flat surfaces, in low-lying cryogenic freezers; and the customary rows of various test tubes and strips of PCR tubes, computers, and the barrage of power cords.

A low-groaning din accented the reality of some pretty sophisticated science taking place, Lev recognized at once. This was no undergraduate playground for interns and adventurous students, but the real deal. It meant power, money, and peer-reviewed propensities for opportunism, in a nutshell.

"Not too shabby," Lev concluded after the three-minute tour. Taman held the Barebones Forest Lantern to get them to and from the concrete lab and sleeping quarters, since the outdoor lights were always turned off.

"There," Taman pointed. "Our infrared surveillance. I'd seen some Big Foot show on Animal Planet and found out what kind of cameras they'd used. And others for a special on woolly mammoths and time-lapse video of changing tundra conditions on Wrangel Island. We modified the detection zones, eliminated incandescent flashes, and added triggers for more images."

"I saw that same show. I did! I was especially interested in the bit about the eyelashes, how large they were, as in the case of goats, to protect the eyes from the stinging ice-laden storms. We humans don't have that advantage."

Taman pointed out the DVREye video system, as it was known, a DigitalEye still photo camera, and the Raptor Cellular system, each among the best available in Belarus.

"Without eBay we wouldn't have had the budget."

Lev noted cameras pinioned in several spots, and Taman assured him there were others that could not be easily detected by any unwanted visitors, including the "black box" camera, a fake.

"What about the normal GPS signals?" Lev enquired.

"I deactivated that component, naturally."

A room adjoining the bathroom had outdoor gear, including the ski poles, anoraks, snow shoes. Cross-country skis were not feasible, given the bogs, Taman had explained. A mudroom. Freshly stored coverlets, parkas, hanging mittens, all drying from recent forays out into the wild.

Lev had his own room adjoining a common area where bookcases teemed with dog-eared scientific journals and books.

"How long are we here for?" the professor asked. "I assume we are spending the night indoors?"

"Yes," Taman said. "I hardly see any alternative." They both could hear the pounding through the forest. The storm was intensifying by the minute.

"*Prahnoz nie jość dobra. Jany pradkazvajuć try futa sniehu na praciahu Hrodna , boĺš u Vaŭkavysku. Jak dva hady tamu,*" Ulyana interjected, as she headed into the kitchen to prepare tea for the three of them.

Taman translated. A bad forecast. A lot of snow predicted, like two years before.

"What happened two years ago?"

"We got slammed," Taman said. "Cyclone Javier dumped almost a foot of snow on Minsk in one day – that would be thought of us as mild, out here, but up in Minsk – a big deal, with wind speeds of 75 feet per second, about the speed of a hare escaping a wolf. There were traffic jams. A few people died. But here in parts of the forest, the snow depths hit between six and ten feet, the winds were of hurricane force, about 110 miles per hour. We had damage to the lab. Out there, well, worse, but of course trees do offer varieties of protection."

Lev looked quizzically at Taman. The expression needed no clarification: *How could my brother have survived out there?*

"Tomorrow morning, I suggest we discuss everything."

"Can I get e-mail here?"

"Use my computer," Taman said, "but only between 3:30 am and 3:32 am. I suspect, from my few early years in the military, that's the window when they change shifts."

"They?"

"They."

At 3:31 am, Lev was awake, seated at the computer, and sent two e-mails, one to Sasha – "I'm here, it starts soon, I love you," and the other to Malcolm – "Mal, it's urgent, make sure I signed off that Shibboleth thing properly. Check that there are no images, no info, nothing. Love to you and yours."

Chapter 65
Visions of the Shtetel

The night brought it all back to Lev, that and the wind, the snow, a tenebrous pall reminiscent, somehow, of that *Kehilla Kedosha*, a congregation sanctified by the Children of Israel, no fiddlers swaying atop the few slippery brick roofs of two brothers' *shtetel*, exactly, but sentiments that certainly comported with something like that.

A small pre-school, far shule, where David had then moved on to the *yeshiva* in a nearby village, but Simon was on his way to the local *cheder* for children; built of wood, this enclave of Ashkenazim (European) learning would be aromatically enveloped in mlinchkes of cheese, dishes of dried fruit, lokshen and miltz, noodles and spleen, welcome dishes, especially with raw onions added, to one who was perpetually hungry. And daily study of the Old Testament.

For the older members of the community, it was a world of marriages and evening walks along the cemetery, whose denizens were invited to weddings; where scholarship served God, the Torah, the Talmud and the learned *rebbes* as much as the lowly peddlers with their enchanted witticisms.

Ropshitzer gabbai. That was the role of the one distinctive cousin he could presently recall. His cousin Morris, whose role it was to serve as humble assistant to the *rebbe* of his village, taking and receiving slips of paper. Morris was one of many beadles, or secretaries. Such duties could be summarized in a name, a useless name, now, certainly to Lev – there were many such words – dating back millennia throughout Jewish history. And what was written on those messages? Morris was dead, unquestionably a victim of the Holocaust. No one but Lev would know he had ever lived.

And perhaps Simon.

There was no crime, Lev remembered – only rituals, covenants (dating back to Abraham), prayers morning, noon, and night, the ever-whispered wisdom of Maimonides as against the everyday hardship of his fellow *proste yiden*, those with little education, and the patient Messiah, for whom *haskalah*, the Jewish Enlightenment, and Zionism were not even words, but socialist concepts, as much as the work of Lenin for the Jews in London and Dublin.

Theodor Herzl's book, *Alteneuland*, would stir their dreams, but – as he discovered in high school, far removed from "Poilin" ("Here shall we spend the night," as it was translated), Lev had read the story of little people, "Kleine Menschelech" in the great novel – *Dos Kleine Menshele* ("The Little Man") in 1863, by Mendele Moycher Seforim (Shalom Yakov Abramovitch), considered to have been written by Solomon Naumovich Rabinovic, under the pen name Sholem Aleichem, the first great work of fiction in modern Yiddish literature – four authors in one, according to legend.

Lev remembered so many of the old black and white photographs from a book that Sasha's aunt had sent to his book-adoring father-in-law. It was entitled *The Old Country*, by Abraham Shulam and with a foreword by Isaac Bashevis Singer, published by Charles Scribner's Sons in New York around 1974, the year Lev first journeyed to the Kingdom of Bhutan, followed by a mad impulsive dash to newly independent Bangladesh, of all places. His aunt had written, "Mom, Dad, I thought you'd like to see it. Love, Thelma."

Of course, she knew her parents were long dead. In the worst sort of way. That there was no way Lev could forward the book on, with its amazing images of a former world. That world had all but died, save for such surreal memories. Resonating survivors abutted up against the truncations necessary to the published page. But she wanted Lev and Sasha to see it.

Of "children fetching water" (the first image in the book); of the entire community of elders posing for a photograph on the day a group of their friends was departing for America (huge!); of a frowning kid selling Bialys on a cobblestone street in Warsaw, and so forth. Priceless imagery. But *The Old Country* was a divining rod for the adult Lev, which was what his sister-in-law had apparently intended it to be. That it had been conveyed within the family, well, there was no other source to turn to, in his case.

In retrospect, at this moment, he could not be more grateful.

Lev, despite having tried to renounce his connections, feelings, even memories of all that had been his childhood, was slavishly devoted to this book, for it put faces and memories on lost souls and vacancies within the legends of his own upbringing. He kept it near his beside for many decades, the way some people stash chocolate-covered Himalayan hemp seeds nearby in case of an emergency.

He recalled all the wonderful solemnity of those seated in the old *shuls*, and the stern but melancholic gazes into eternity of a *cheder* in Vilna, where so many of his relatives had come from. Of *yentas* and alfalfa-trundlers; Bolshevik boot-makers who had read up on Marx and others, and women with bright bunches of yellow roses in spring – the Catholics who would buy those intoxicating flowers loved the yellow ones especially, he recalled; and every conceivable handyman, carpenter, grime upon grime in compositions leavened as by a Rembrandt, long-lasting scenes and faces as wrinkled as any land Lev had gazed upon, wondering about each and every one of them, all relatives, the way, somewhere, James Joyce or Blaise Pascal, or both, had once written that "chance" is what guided them throughout their lives.

Lapidary clarity. Such that wherever he looked down, and found something, or felt its intimation, it was precisely what he'd been searching for. It had been there all along. Not just something, but someone.

All of these people were related, engraved in the hardwiring of twelve tribes of intertwining divine and desperate heaviness. Like a Bach Passacaglia and Fugue, sobering harmony heralding some triumphant return known only to that cult of destiny David Lev had long ago escaped. Yet a distant future now lay low upon him. There was no denying it. Destiny had, in fact, outsmarted the brainy ecologist. He knew he was, to put it politely, fucked.

These memories came from the world his childhood had known. And if he had known them – with his bit of learning and preparations for a grand departure to Dublin – what about the much younger Simon? What could the little boy have grasped of the daily mysticism? He was a child, surely oblivious of those vagaries and looming importunities of survival. The Judaic equivalent of "Ave Maria" where there was no one left in the forest to hear the falling of a tree or remember any of the merriment. Of those many elegant and beautiful girls and boys that were no more.

Except the one vague clamor of a possibility haunting the entire scenario.

Chapter 66
Twelve Feet of Snow

One man outside, as the onslaught of spring had come upon that forest with darkness collapsing, not descending, as the full-bellowing squall slammed them. A ferocious cacophony of whitening downpour that seemed directed everywhere, as if ball lightning had turned electricity into tidal waves comprised of frantic snowballs, a wild frisson of amps with no peripheries.

"Make sure you keep your iPhone off," said Taman. "It shouldn't work anyway. No apps, no electronic devices."

"Got it," Lev replied. "What about my pacemaker?"

"You're joking?"

"I never mentioned it to you?"

"No. It may be a soft target, so to speak. That's something I never thought about. Titanium prosthetics, definitely a problem."

"I'm lucky in that respect. We both are."

"Right."

Enormous limbs of trees broke off, the roof above yawned and seemed to cry out with the stress of so much happening overhead. A very different scenario from what conservationists, particularly students of phenology, knew to be "spring forcing."

Taman and his assistant, Ulyana, had taken leave to their private little rooms, as if unconcerned, while Lev lay awake on his cot, covered in three heavy and uncomfortable woolen blankets, thinking back over an abyss he had all but dismembered with the perverse luxury of time, waiting for a limb to crash down upon them all – for the limbo of this moment in his life to swallow him whole.

The snow seemed to eclipse not only the memory of names and villages, but the entirety of the continent.

The snowstorm was telling, and of such ferocity as to be endemic to East European memories. There were no ethics in such a storm. Only flight. To flee the worst of it, if you could manage to do so. Not possible without some form of accomplice or rope tied around your waist that was sure to guide you back from the outhouse to the main house.

Take shelter in any opportunity that manifested with fleece, hair, anything to batten down the hatches of exposure in one's being, even the crudest piece of crumbling or cardboard wall, in whatever configuration. Even a hole in the ground was preferable to such indescribable cold.

The dead, Lev remembered from one day back then, seemed to live in a grandiose style, compared with those confronted with the fierce bitterness of East European winters. Such contradictions were not easily reconciled with the continuing confusion that spread plague-like over the minds of children cast backwards and upside down into the history that was their discord of the mid-1930s. Those same dead would be invited to Purim celebrations in the old days. It was the custom, even, to fraternize around cemeteries. The invitation was always open.

Now branches began to hit the roofs of the compound and seemed to call for emergency responses, but Taman and his assistant were silent, no contrition in light of the priority that was a good night's sleep. No one stirred, only David Lev's mired reminiscences and present sense of utter fear: How could he have arrived at this point in time, with an alleged brother somewhere out there, in this calamity of intersecting centuries, nation-states, ferocities of nature and of human folly? The same tiresome conundrum plagued him this night.

The sheer chimera of biological survival against the most daunting odds in vertebrate history. Wherein, thought Lev, remembering from long before, in some awkward classroom, or alone in a bedroom safe and secure, Shakespeare's "Coriolanus," the lines, or to the effect, as spoken, if he recalled, to his friend, Menenius Agrippa... *Within thine eyes sat twenty thousand deaths, In thy hands clutch'd many millions in Thy lying tongue, both numbers...* well into Act Three, Scene Three.

And if it were a truth, then – should he remember other than Yiddish a language in which to speak, certainly not Hebrew, adopted by the State of Israel, and surely no English – he would likely declare with adamantine resolve that every day he survived was another conquest against the unthinkable.

Unthinkable, but determinable: if, indeed, Simon was out there, had been out there all this time... the odds against it were unfathomable, and Lev thought about the Nazis dying by the thousands in the deep Russian freeze of 1941. Good for them, the legends of frozen eyes, of grease freezing inside their munitions.

The snow continued to fall.

Simon himself, a living verity amid those who hated the Simons of the world.

Throughout the millions of years of hominid phylogeny, no other primate had ever committed a Holocaust. Chimpanzees, and a few of the other 650-odd primate species, might be violent, kill, even resort to infanticide on the rare occasion, but nothing like... Germany, with its Austrian, Polish, Hungarian, Italian, French, and other accomplices in the unimaginable action against the Jews.

Take hominid evolution even further, to the orangutans with whom Lev had spent time in Borneo, not a bad bone in their bodies. They might be a hair more genetically separated from humans than chimpanzees, but they were closer to us emotionally. They did not cannibalize.

And he looked out into the blizzard and wondered how, if it were indeed true, his younger brother had fought to prove, day by day, and every night, that to live was to be victorious; above the heads of those who had sought to slaughter him, and had been missed, perhaps by a stray bullet, a blink of the eye of a soldier standing guard, or in pursuit of a child who ran off into the forest, never to be found. And those who might notice, who would surely, silently notice, they had been slaughtered.

It all occurred in some singular moment of distraction on the part of a soldier, who put out his cigarette, decided the chase was too tiresome a notion, figured he had better things to do, a whole community of Jews – and some Catholics – to *tidy up*.

A child. Surviving high up in a tree? Twelve-foot snow drifts in one night? All of it impossible, mused Lev.

But a café in Rio blown up that no one was talking about? A cover-up? By whom? For what reason? Despite Taman's frankness, there remained amid that multiplicity of motives and harm's way, the undeviating fact of Simon.

David Lev was here, in southwestern Belarus, on the Polish border, in the late terrible spring, to find his lost brother. On a tourist visa. With not even the slightest clue as to what he would do, should the two siblings be reunited. What did that even mean? Duties, responsibilities, ethical quandaries unlike any Lev had ever contemplated.

But the culmination of all such speculative morality and human catalysis was the fact of Sarah's hand-written account.

Chapter 67
The Diary

A Xerox copy of the diary was neatly wrapped in butcher paper and cosseted beneath a sweater in Taman's backpack, from which he now, appearing before the exhausted Lev, removed with extra care, noting that his assistant Ulyana had gone outside with a broom, stepladder, hammer, lantern, and pliers to fix three malfunctioning surveillance cameras and to assess the damage to their scientific station from the gale force winds heavy with sleet and the weight of continuing wet snow.

Taman hid the diary when he saw Ulyana suddenly returning, clearly winded.

Lev realized at once the importance to Taman, to Lev himself, that the existence of this presumably salient and revelatory – if not incriminating – document remain hidden to all other eyes.

"*Ja dumaju, što dvanaccać futaŭ u niekatorych miescach. Dziŭna, ci nie tak?*"

"What did she say?" Lev asked.

Taman explained that, "Ulyana believes it may have snowed twelve feet so far in some places in the forest."

"That's not possible in one night."

"The drifts. Yes. It is possible."

Ulyana poured herself a cup of coffee, drank it in all of three swigs, put down the glass, and took hold, in the mudroom, of a larger shovel, then exited the compound.

Taman, certain she was back outside, reached for the more than seventy-year-old faded book, whose entries he had translated, and placed the folder on the table before them.

Lev stared at it.

It had been a long night. He'd only managed to pass out sometime just before the deeply divisive sunrise that had come upon the world bearing several feet of icy thick mist, wind, even the sound of that ball-like lightning amid the mini-tornados sweeping the unseen barriers beyond. Blasting debris of biomass in a cyclone, one might think, of kinetic force for which there could be no easy explanations if one had tried to mount a theory of human survival at the brunt of such a storm, Lev's obsession at this point in considering the plight of Simon – out there.

If meteorology were morphing in the twenty-first century, a function of climate change, then they were in the front trenches. Of that Lev was certain, at the apparent heart of such a storm, a storm that was continuing into daylight.

"And we have not seen the end of it," Taman asserted, recognizing the utter inexplicability in Lev's questioning eyes.

"We had late winter storms similar to this last year," he added. "Climate disruptions, most certainly. But this is worse."

"Disruptions? No. This is the 600,000-year-old new norm. Polar bears will be seeking refuge in Belarus before long."

Taman checked the barometric readings, then switched on the station radio to the national weather channel to determine whether his compound's barograph was accurate. Minsk had received some two feet of snow, ten-foot drifts, sustaining 40 mile per hour winds. Conditions were continuing to deteriorate.

"That would explain 956.3123 millibars of mercury," he said as if by rote. "Of course, we've had much worse."

Lev was stuck on 956.

"In inches. We use inches."

"That's right. Slightly old-fashioned." Taman was quick to convert it with his Android.

"28.24," I think.

Ulyana could be seen, suddenly, moving past a window outside, shovel and broom amassed in an effort to break the weight of some of the more enormous icicles bearing down on the gutters.

"You obviously don't trust her?"

"Why would I trust her? No," Taman said somewhat adamantly. "Trust went out the window in Rio."

"Does she know what happened?"

"She must have heard something. Keep in mind she is employed by the government. Of course, there was never anything on the Internet, or in the papers here. It's as if nothing ever happened."

"I know. As I'm sure I mentioned, it was the same in the States. I almost imagined I could let it fade from memory, aside from the bursitis in my left shoulder."

"For me, it was only a matter of time before they made their next move."

"They?"

"This is Belarus. Anybody. I think I've clarified the many motives at play, so to speak."

"So to speak. Yes, well. Does she have a gun?"

"Ulyana? Let's just say, it's not allowed. Although nearly two percent of all Belarusians have guns. So, probably, yes."

His voice went closer to a whisper: "I keep a handgun. A 9mm Makarov pistol, made by the Soviets. I obtained it after you and I first met. But you never heard that. And it is not registered."

"I should have an idea where you keep it."

"It's always in my pack whenever I go out there. I pray I never have to use it."

Taman had been living with thoughts of this for over three years. Not unlike Lev, he, too, was pitched headlong against unknown cavities in the fabric of what had been a stable existence, pivoting in a heavily militarized zone that was his homeland, and especially so his research in this forest.

But he was also as inured to the physiological plight of the one unclear mammal out there: the survivor. And was trapped in a kind of bizarre guilt that encompassed his own roots; the many deaths; of not being in any kind of position to jump in. The whole situation teetered for him personally on the entire disgrace of World War II. Taman was implicated in ways that left him disoriented, with no clear exit strategy. He truly did not know which motives were swirling across his turf, his personage, his career, the safety of his family. All was blurred and could not be rectified. Or at least there was no hope of it until now.

That need for clarity and resolve drove him to seek out the professor.

He'd been living amid a constellation of fears. Seeing in every single daily or weekly irregularity the beginning of the end. No way to live. Anna, Taman's mother, had warned him: "*Vy pavinny zviarnucca da psichaanalityka*" (You need to see a shrink).

***

Lev picked up the diary. "So this is it?"

"Yes," Taman replied solemnly.

There were some six pages spanning many years, with the paper, although a copy, showing signs of stain, rain having mottled some of the material.

"These are your notes?"

"As precise a translation as I could. Maybe some mistakes. You'll forgive my English. She wrote in both Belarusian – the Medieval Lacinka script, partly Latin – and Polish. After the Soviets came in, she adopted Russian as a matter of survival."

"And you're fluent in all three languages?"

"Yes, of course. Here, unlike the complaining character in Chekhov's "Three Sisters," knowing many languages proves useful."

"Presumably the original is safely hidden?"

"Stashed where nobody would ever think to look. I've made no digital copy."

Lev paused before starting to read, because there before him was a storm. And out there, in the storm, his own kin. The juxtaposition left him struggling for oxygen.

He said, "You conveyed the urgency. And I now appreciate the fact that in a way, I guess, I could not have before this moment. You've been out there for how many years?"

"Almost five."

"Five. At night?"

"Of course. With the proper gear. It is easier than during summer in some respects, given the number of ticks, chiggers, and other stinging, biting, poisonous organisms."

"Tomorrow, assuming the storm lets up a bit, we'll go out?"

"Definitely. If you are ready. I know you've pictured the encounter in your mind."

"A hundred times," sighed Lev.

"I mean, if we go out there and find him, which we probably will. What then? What do you propose? How is he likely to react to seeing you? You to him?"

"He's not a guinea pig for the worst century in human history, if that's what you mean. I won't permit that." Lev was feeling truly hesitant about beginning to read the diary.

"Do you believe in justice?" Taman asked him.

"It would not be just to parade my brother around like some freak of nature, driven by a chauffeur to speaking engagements for Jewish organizations, universities, the World Court. An Ivy League university commencement address. He's not the Elephant Man. And he's not Ishi."

"Ishi?"

"An American Indian, the last of his tribe. Like Vladimir Arsenyev's story of Dersu Uzala. Ishi ended up at a university, not quite a zoo animal – he retained his dignity, under the circumstances, and given the enormous compassion of the person and institution that helped him – but, still, a kind of ethnographic mascot, an anachronism in the twentieth century. I'll read it now."

Lev began slowly, a pointed flashlight and his index finger abetting the arduous journey:

I sensed it coming first with the wood merchants of Hajnowka who stayed clear of the Rezerwat. Some mentioned the signs in the park had changed to German. The Russian, the Polish, scratched out.

No one had reported seeing even a single Zubr in months.

We all knew what the locals were saying. A team headed by the Reichsmarschall had arrived. Someone heard they were attached to the zoo in Berlin.

Father mentioned Czechoslavakia falling. Nobody could understand. Where were the Americans, the British, the French? The long-standing Treaty of Versailles? Then Norway was attacked, a treaty with the Soviet Union failing to stop one thousand German tanks, what Father termed "Panzer Divisions."

Poland's leaders have fled the country. Russians are showing up from the eastern side.

Lev put down Taman's notes, removing his glasses. "I'm confused."

"What?"

"She was just a kid. This is heady stuff for a kid."

"Yes. Precocious, no doubt about it. But you'll soon see that some of it was written in the aftermath, some, maybe, in real time. I don't know. I think you are right, though. She must have reconstructed some of her earliest reminiscences at the time of the Nuremburg Trials. Somebody might have coached her to do so, thinking there could be reparations, I'm not sure. But there is very little context or time-frame. Just a couple of things. It is perplexing. I struggle with the pronouns when translating. First person, second person, third person. My English is not bad, I think –"

"You're fluent."

"Well, maybe, but this was strange. *'Poland's leaders have fled* - бежали, not had fled, бежал... *Russians are showing up...'* Present tense, no?"

"Yes. But, as you say, if she was older, during, or after the Nuremburg Trials, couldn't she have gone in search of Simon? She probably knew the forest well. I would have thought that –"

"That what? No way, David. After the War, Poland was still engaged in systematic anti-Semitic reprisals. Everyone blamed the Jews for everything. Surely you are aware of that. For the Germans, for the Russians, even the fact that the Germans got the most benefit of the Marshall Plan. There is evidence of multiple killings. Sarah would have been in her late teens. On the eve of marriage, children. No one was allowed into this forest. Everyone was covering their tracks."

Lev resumed reading.

I think your uncle worked either on Stoczek or Waszkiewcza Ulitza, at least for several months, going, incredibly, in someone's automobile. I don't remember how that worked, with the RFA, the Reichsforstamt monsters evidently roaming the forests, from the Bezirk Bialystock forest preserves to Bialowieza.

The Bandenkekämpfung I think it was called - German, what a language - they were gangs roving in the woods, the bogs, the fields. And then the day came, we all heard about it: the Gaujägermeister. Master hunter. And the first mention of the name of Hermann Wilhelm Göring, they call him the President of the Reichstag, and something named the German Falcon Order. The bastards have inundated everything.

Now it is a fact, we cannot deny it: all these terrible names descending upon us in the newspapers, in the night, in the mornings - Lutfwaffe, the Jagdkommandos, or Jakados, Austrian hunters brought in by the Nazis. They are hunting Jews in the forest.

Today word has come that it is possible a few of the Nazis have actually eaten body parts of the Jews. Cannibalized them. Everyone will deny it. History will deny it. Germany will never accept such a thing. But that is what I have heard whispered in hidden corners. Opinions are flying in every direction, as the invasion and atrocities intensify.

We thought we knew the Germans. But we know only too well about the Panzers. The very same ones, we are told that enforced, then burned down the ghetto in Warsaw. But if this business of cannibalism is even remotely true, then surely the world has come to an end.

"I'd heard about cannibalism in some of the camps."

"No, I never heard it."

Lev re-aimed his flashlight, his thoughts quaking.

After the war I searched for him at the Jewish cemetery on Okopowa Street. Nothing. The world is dead.

"So, there it is. She did go seeking him out."

"We have no knowledge of her having ever lived in Warsaw, David. It's a mystery."

We know well that the Blitzkrieg Manifesto, as some are labeling it, was determined as an ideology to void the presence of the Jews. Of all mammals hiding out in the forest as well. It would explain the presence early on of the zookeepers from Berlin. I have no idea what backbreeding means, but that was their plan. I have it on authority.

On our side, Anixt was annihilated. Twenty-thousand Jews shot in the forest. What is there to say? The Wehrmact occupiers retreated in July of 1944, destroying even Göring's hunting lodge in the forest. So much for the legacy, on the Polish side, of Sigismund the Old, who had endeavored to protect the poor animals.

Everything that has come out. It all goes to that day I saw with my own eyes. Place-names of the unimaginably grotesque like Majdanek, Belzec, Sobibor, and Auschwitz, also Chelmno and, of course, closest to home, Treblinka. The Allied Forces entered the camps and brought out the photographs. And I've heard it lamented by those who should know better: Oh, it can't be true, it can't be possible, they keep saying. Or, We never knew. But we now know that it is true. Neighbors turning on neighbors. That these fuckers were lying to themselves, lying to history.

All of the region marked by our great forests, and our once serene Narewka River, and those adjacent country shops in such picturesque places as Gródek and Budy and Teremiski - remember Teremiski? Help thy neighbor? A fresh loaf of bread, a slice of cheese - everyone was so kind. Life was normal. Now, who is kidding whom?

Taman acknowledged with love, "She simply adored the word 'fucker' even in the nursing home. I thought it was on account of, you know, the Alzheimer's. No. Her whole experience, it was clearly enunciated in that blitzkrieg of a word. A poetic oath, as solemn, lyrical, and truthful as the German language she so detested, notwithstanding a few lovely books by Rilke that my Sofia had given her and which, in German, she had devoured, in the early months of her being under so-called "protected adult" status at the nursing home. And I understand."

I went, one day during 1953, June 27th, you'll appreciate that, to the old brick building, amid butterflies and mosquitoes. It was summertime, of course, along General Waszkiewicza Street. As a child, it was named Stoczek Street.

"The second time she's mentioned that street."

"Yeah," Taman acknowledged. "It was obviously very important to her."

By 1953 there was a Star of David on the building. I'm glad we got out to the Belarus side. The coffee is better. There were a variety of tarts fashioned with sweetened egg yoke and served on pineapple buns, Szarlotka cake with our tiny apples, unthinkable during the War. Let alone halva, pumpernickel, kasha, strudels, mushroom soup, homentashen, barley preparations, home-cooked soft muffins. How are you doing, Simon? Insane question, I know. But is it possible that you might still be out there?

"That's all Jewish food," David said, underscoring the obvious.
"I know."
"And June 27th, 1953?"
"No idea," said Taman.

Stahlecker's, January 31, 1942. Report regarding killings of the Einsatzgruppen: Of the more than 200,000 Jews in Vilnius, the vast majority were executed. In Simon's parents' villages, all were machine-gunned, stabbed, raped, shovels to the head, those that were still breathing. In Anikst, a 'Who's Who' of Nazis, I think they were part of something called Reichskommissiatostan. Their goal was to kill all the Jews. Why Jews, I always wondered. Still I wonder. It was easy for the Nazis. They had occupied Vilnius since the early 1930s. Information was not information, but rumor. Rumor was hushed. Even that which was hushed, was hushed up. Nobody knew. I didn't know.

The railroad tracks ran nearby. We all knew that. And the Jews, like the wisent, were transported - not in steerage, but like freight. Hardly any air, no water. People vomiting on one another. Crying. Starving. Freezing. The conditions were described by one little boy, your same age, Simon, who also got away, somehow impossibly; he managed to squeeze out of one of those trains, he ran, they never caught him, as far as I know.

"Of course! My God. June 27th was Simon's birthday!" Lev's mind was racing. "These street names. And the town, Anikst, not Anixt, or not that I have ever seen it written that way, in Lithuanian. It's now known as Anyksciaia, or Onikshty, however the hell one pronounces it, near Ukmerge Uyezd, or Vilkomir, where my people came from on both Grandma's and Grandpa's sides of the family."

Lev was dizzy. General Waszkiewicza Street. He vaguely remembered the synagogue. But it was not a fancy synagogue. A *shul*, yes. But of wood – or at least his memory brought up wood. Not brick. Or maybe he was confusing it with a town a hundred miles away. Nothing was in sharp focus.

"Drink some tea," Taman appealed, offering a fresh mug.

"Thanks, no. I just have to get through this." His heart, racing. He could hear the pounding of the blood in his ears.

You were among those I saw, Simon. I saw you that morning with the chicken, the bullet that ricocheted into your left eye. The blood squirted onto others. Rounded up, I heard them order you naked, every one of you, your mother, your father, even the Rebbe, and then marched into the woods. I was frozen with fear. There was nothing I could do, two stories above the street, hidden behind the curtains. You'll remember the room. I was peeking. My parents were horrified. We were all horrified. But it was join you, die, or be silent. My parents' insistence on silence. I'm so sorry, Simon. What would you have had me do?

"I had trouble with that. The tenses. English past perfect can be insanely difficult to make sense of."

Lev kept reading:

A little girl. I knew everyone who helped them. I could have spelled every one of their names at the Nuremburg Trials. But I was not invited to Nuremburg nor fancied going there. All of our school friends – Kacper, Adam, Tomacz, Kamila, Jacek, Piotrek, Martin – remember little Martin's mom, a Jew-hater like few in the town, and of course Bogdan, I could have predicted that; and Patrycja, Wojciech – their fathers were summoned by the Nazis who had strode through our village like kings. Did they hesitate to assist? You tell me, Simon.

Lev was trembling, his breathing at a falter. He could hear his heart now with a kind of end-of-life signal. He wasn't traveling with his blood-pressure cuff. Too heavy, too predictable. Every one of those names he remembered. Each face he could picture. And their parents. Kacper, Adam...

It was not some remote sensing canvas of pixels, or sepia haze of evocative points on a compass. This was black and white and forbidden. He had never seen it before this moment. All of their parents, who had lived next door, down the road, behind the shul; at the bakery, the well, the one hotel in town... He looked at Taman with utter desperation. "Your great-aunt wrote all this? So let me understand what's happening here," and he proceeded, albeit with no singularity in mind, to apply some ancient telescope with nobody at the helm, no sense of honor that was at the beginning of this world, incapable of pushing boundaries, a muscularity of the language, the thinking process he had not had to go near, until now. All these years. Even in Ireland. He had been liberated from the requirements of thinking about his Jewishness. His childhood. The sequence of events leading inexorably, beyond all comprehension, to those two words: The. Holocaust.

And what his involvement might mean in terms of having survived, gotten away – why did he get away, when so many, most, did not?

What right had he, what chutzpa, to be there, safely secured in a research lab, etc., when... when even an old woman in a nursing home had owned up without a moment's reserve... when he, who had had a lifetime to become engaged, viscerally, verbally, even scientifically; to move some semblance of a behemoth forward, away from the same-old horror story of racism, fascism, ethnic cleansing, of Treblinka but, without so much as a thought, even the odd respite, had done absolutely nothing? Voted, of course. Gone to the odd Passover service over the years. Yom Kippur. A deli with a few Jewish friends, though not necessarily Jewish, for

Rosh Hashanah. He belonged to no temple. Now he despised what he had become: assimilated.

He was precisely the version of anonymity, that very personage that Elias Canetti so precisely conceived with the publication in 1960 of his *Masse und Macht*, Crowds and Power. A fellow, picked randomly, who would join the mob rather than resist terror. Lev, by his childish good luck, had been condemned. He could not escape the narrow back alley of a life, no matter how successful every other part might be.

His age was his worst enemy, no defense against it, but that was no excuse: success at growing old, with each day a repetition of the same disgrace, to be alive in the face of everything that had happened. How dare he survive. What right did he have to smile? Hedonism to have married, sheer egotism to have embarked upon a scientific career. As he recalled in a line by the poet Byron:, it was his fatality to live - or words to that effect

If there was any good news for him, it was his willingness not to have children. But even that, if the truth be known, had its biological ambiguities woven into the fabric of official explanations. Often, he lay awake at night and panicked at the thought that he had so let down his parents' legacy through abnegation, blocking the fertility of an ideal – that of his parent's lasting hope in a future, any future, the future of Abraham, Isaac, and Jacob.

Did he stand for anything good, honest, trustworthy, eternal? Were his virtues intact, his character worth mentioning?

Had he done anything, anything whatsoever to make good on the promise of his so-called profession of nature?

Nature lover? This was the worst of all the bullshit. Nature and Hitler. How could English, or any language, allow for both words?

...*fuck*.... he thought. I am dead. And his heart beat faster and faster. And then, without the least resolve – how could any resolve ever emerge? – he dipped back into the imponderable, reflex-ridden pages before him. It was not just a chore, a duty, but a gang-plank. He had lived on it all his life, oblivious to its meaning. Now the words and their entanglements had necessarily caught up with him.

Later, word got out - from those who went with the park authorities and saw long trenches but dared not to meddle with the bogs, the so-called "ancient relics" - I could surmise what must have happened. You, with your cheeky grin, that flopping bundle of curls, bright red, nobody had ever seen such hair - maybe your older brother, I don't remember -

"Older brother!"

"Yes." Taman had sat listening with nerves of not quite steel, as Lev had read aloud, his voice quavering at times, at others, resolved to the indisputable truths. But now, this reference to himself, it stopped Lev as sure as that explosion in Rio.

"I must have met her, then?"

"Yes. At least once or twice," Taman volunteered. "No question about it."

"So. It's all true."

"What did you think?"
"Now I understand. Everything."
"I'm so sorry, David."

...and your way of bouncing around, racing like an Olympian, you would have seen every one of them. Stared at them with absolute confidence. No fear in your one remaining eye. I am certain. You would have gotten away, if anybody could. Like the boy from the train. Yes, I am sure of it. Even in my darkness, I know that you are safe. Free. Forever liberated. I was two miles and fifteen years away, by then. I was never to be free again.

"What's the big blank?" Taman had left most of an open page.
"Keep reading. She must have put the diary away for many years. Then resumed."

The Americans have gone and landed a man on the moon. I know. It sounds crazy. Maybe they sense another Hitler emerging, one who is ready to blow up the world.

I've gone and gotten married, Simon. To a Jew. I want a son who will grow up to be a protester. A historian who loves the forest. Someone of substance. To learn that song of yours.

December 25th. Night. 1967. I am not celebrating, trust me. Although the bells have been ringing and last night all the dogs in the neighborhood were howling alongside the carolers. Whenever I have red chrain sauce and currants along with bialy stuffed with onion, the cabbage well-cooked and rolled with chopped beets and arsenic-free rice; buckwheat pancakes and poppy seed baranki served up with curd cake, and my Mom still saying, "Eat, eat up," I think of you, Simon, out there. My Simon. I never had that son. I was deemed infertile. I don't mind, though. And my husband, he died. Cancer. But it was rapid. So my life goes on.

What could you be eating, Simon? Maybe those mushrooms I remember you were so fond of. When we used to steal away together into the forest? You showed me things no child your age should have known so much about. How did you come by that knowledge? Did you learn it at the shul, the cheder? In the Talmud are there mushrooms? You never told me. But you knew not just the mushrooms, but the bison. The birds. You loved each one of them. You used to sing to them. And they to you. I remember. The crickets. Even the strange living plants on the trees. You said they could sing as well. I thought, were you mad? Of course you were mad. At age five we were all mad. You had a lovely voice. The voice of a born musician. You were a little boy. The most beautiful little boy. How did you know so much? So many bird songs?

I remember the Rebbe called you Orfeo. Arfiej in Belarusian. Only many years later did I remember that, and found out about it, the famous character in Italian, and before that, Greek mythology. I still remember a theme. Yes, truly I can hum it. And some nights, like this one, I hum it to myself, Simon. To forget. To believe. To hope. It is the prayer you gave to me. It is ancient music and it calms my heart to a still; but I was too young to ask how you could possibly have come by it. Or did I? So much has happened, maybe I forgot. But I do remember how we chatted away like it was the perfect paradise all around us. That forest which surrounded us. The blessings of childhood. I miss them. I miss you.

Late night. February. 1968. Simon, now, you will be happy to know I have two pet cats. Alley cats, I suppose. Madejówna and Ziębowie I call them. And a third one, lingering feral traits, who now keeps their company slavishly. Arkadiusz. His tail has been eaten off by something ferocious out there. Perhaps a fox. But still, in winter, sometimes all three venture out at night together, I have no idea why – it is freezing cold, as you would remember well, and still they prefer to wander. But where do they go? Can it be that important a place to brave such weather? What do they do? How do they survive the cold? It is something terrible ways. Nature can be something other than your magical songs.

It is horrible these days, we are all so poor, toilet paper is at a premium. The Communists, and to the southeast, where we grew up, there is talk of socialism, of breaking away from the USSR. But how? There is sure to be another war. When my mind goes back... to when we played together in the forest. But I must stop this.

"You wrote this?"
"I did my best to translate."
"Your great-aunt wrote like a poet."
"She was a housewife, well read."

December 8, 1991. They are singing "My Belarusy" in the streets. But all this talk of freedom, independence... Are you eating Zubrówka, getting drunk, I pray? I will always pray for you, Simon.

The diary notes ended there.

David mulled over everything in this strange epistle, grasping at loss across decades, emerging from some centrality of a little girl's heart, bespoken over years, as that girl grew up, gained perspective, her historic observations turning to revulsion and incredulity, then, ultimately a nostalgia, on an ascending note, for the long ago. For her Orféo.

Taman nodded with a look of deep commiseration. Lev was too spellbound with lament, overcome with a benumbing to rationally respond or even formulate the least summation. It would take him more than a response to ever truly respond.

"Zubrówka. I think I remember that word," Lev ventured with a strange and distant speculation, his mind distilled into one colossal and enigmatic teardrop that could not express itself, no way. Ever. They belonged to the Earth, all those tears.

Taman saw it all, but tried to embark on the steadiness they would need soon enough. "We call it bison-grass. *Anthoxanthum nitens*. Perhaps you remember? It's a native species used in making beer. Some have referred to it like manna, sweet and holy grass that they still place before churches on holy days across Europe."

"The beer. That's it! Zubrówka."

Taman handed David the six pages, folded into a letter-sized envelope. "It's yours. Best that you hide it. Keep it close to you."

© M.C. Tobias

Chapter 68
Life and Death in the Forest

"They also use it in making some Polish vodka. A good sweetener. The wisent – well, the females," and Taman labored on, effecting small-talk, in relation to the morbid realities engulfing their communion. But he had to engender this superficial palaver if only to instill some cornerstone of where they were, what they were doing there, amid hostilities and paranoia and the reality that at any moment they could be arrested, shot, the whole house falling down, worlds colliding, an international incident, or worse.

There were drones, even amid these windy weather conditions. Men with hunting hounds and machine guns. After his immortality patent. That was the mushroom Taman had hidden from everyone but Sofia. And she had made light of it, spoken aloud – knowing the world was bugged – "ahh, Micky Mouse!" she profusely exclaimed.

Over four hundred species of lichen out there. And new data on the False Tinder Fungus which, in combination with the known 8,673 species of insects, and counting – all high in protein – and the biomolecule bonanza Taman had been tracking, first with squirrels and the boars, but then the wisent, which led him to Simon and the first photographs he acquired of Simon, and all of the guilt associated with his not having disclosed those images to David, out of fear, so much unholy fear.

Because Taman had used Simon, he could not think of his role in any other way, at least initially. Yes, it is true – he often recited to himself – I thought he was a lost tourist, out there for many years, but not, not Simon Lev.

Not a Holocaust survivor. But Taman had actually sampled DNA from human feces he was sure were those of this "tourist" who was clearly spending a lot of time at the entrance to a gigantic cavernous hole in an oak tree. Bigger than the Polish oaks, with all their patriotic attributions, names like the early fifteenth-century Polish King Władysław II Jagiełło, or the King of Nieznanowo, The Guardian of Zwierzyniec, the Great Mamamuszi.

Smothering with decay and thereby metabolically inducing the necessary food source and warmth that humanity had fought for over a million years to achieve its *Homo sapiens sapiens* status, or, more precisely, inferiority status. It all now culminated in the survival tactics of a lesser god, who could not fly like the birds, or swim like the fish, or chew cud with anywhere near the expertise of a 2,500-pound wisent.

Taman had experimented, sought information, calculated odds, crawled through icy slime and sucking mud to figure out this insane scenario, all shrouded – with little confidence – in the hoped for secrecy of his own assistants, his grad students. Now it all added up to a sense of horrible shame. And the realization that he had failed at a fundamental level. There were no secrets in this forest.

At 3:30 a.m. Taman checked the multiple pings on his computer.

"There are two e-mails for you, David. Hurry, please."

Lev rushed to Taman's desk. The first e-mail was from Sasha. Reassurances, love and good wishes, as only she could coin phrases that, on e-mail, always seemed trite.

The other was peculiar, from Malcolm: *Prof, did you ever like take note of those spams during Rio Josh had sent? KS, iber azhentsy and Prometej.*

Lev gazed at the words, but he could not remember a damned thing.

Well the word Prometej apparently refers to a very distinguished open library, a fine publisher. Nothing to do with Orwell, the Ukrainian Preface he wrote in that edition to his Animal Farm, unless it's a major leap. It's the other words that Josh retrieved, KS and iber azhentsy that definitely bear some scrutinizing. Ask your buddy, the stalker, but I've done some rummaging about and I think it stands for a consortium, or multinational that's off the grid, but refers to Kiberseedlings, or cyber, either in Russian or Belarusian. Or not. If it's connected to Orwell, and that publisher of the said Ukrainian edition Preface, it was for the DPCs, the Displaced Persons Camps. For their liberation. So you got me, boss. Utterly confusing. Especially the mysterious signature, jv, that is, if it is a signature.

"jv?" Lev pondered it. Nothing came to his purview. He was lost in all this.

"What do you make of this?" Lev asked Taman, who sat beside the professor and read the e-mail off his own computer.

"I don't know of such a company. This, you say, was sent to you when we were in Rio?"

"Apparently. I don't remember."

Taman shut down the computer, then tried charting in his mind the linear timeframes of his own discoveries, and what he might have said to whom. He had endeavored to keep a tight lid, so tight, in fact, he never even mentioned it to his own Sophia.

Taman was charting the linear timeframes of his own discoveries, and what he might have said to whom. He had endeavored to keep a tight lid, so tight, in fact, he never even mentioned it to his own Sophia.

"Do you know somebody with the initials jv?"

"Nobody comes to mind. J equals Jew, maybe, or Jewish? V, V ... I have no idea. A B, well that would be different, perhaps. A Jew in Bialowieza. But V. No. I don't know. Certainly no Victory."

And that's when something sparked in his memory coils... that guy from the Embassy, or Consulate... Vespers. John Vespers. Commercial attaché.

Lev dizzied. "Taman, did you EVER publish anything about Simon?"

"About Simon? Absolutely not. Only on mushroom data. And certain oncologically curative properties, potentially, of the forests of Eastern Europe."

"Of Bialowieza in particular?"

"Yes. With the certainty – after so many years out there – that it was a safe haven. The public is not allowed, David. Only a very intensely vetted short-list of scientists. I can count them on one hand. Myself, three grad students, and Ulyana and her boyfriend, who comes only sporadically."

"Any dealings with anyone over the Internet about this biochemistry, so-called, that you've been studying for five years? Any other Americans, in particular?"

"No. Just one peer-reviewed scientific piece, as I mentioned to you."

"And you have never been to the States?"

"Someday, I hope."

Lev could not be certain, of course, but the coincidence – whatever it meant – was now contaminating every neural second in his universe.

Taman was deeply on the spot. He could not recall, precisely, just how much or how little he had, in fact, divulged to his three graduate students, the one in particular. Even Ulyana could have been much smarter, much more of a sneak than he'd ever imagined her to be.

Demons had invaded his psyche. He was tormented. Feeling very old, wizened, near collapse. He could feel his heavy breathing and imagined a telltale pain in his chest. It was just nerves.

***

At length, Taman admitted, "It's always possible that something got out. Look at you with your graduate student. You even brought him with you to Rio, and you are talking about the extinction of life on earth. I have been, conversely, researching the salvation of life on earth. So we are, both of us professors, with students, colleagues, worlds apart, but connected in the sense that we cannot be one hundred percent sure who is listening, who is reading, who is watching, plagiarizing, outright stealing our data from inside the scientific community, and who is not. I do not claim immunity from – how do you call it, eavesdroppers? Especially on the Web. I work for the State. State-owned universities, utilities, police, military, corporations, embassies. I cannot control that domain of snoops. I cannot presume to control any of it, not ultimately. And like I said, there have been sightings of the rare drone over this forest. Are you surprised? This is the real Belarus that those of us committed to excellence in science have always had to cope with. It may sound strange to you, but we are still, in some respects, in the shadows of the Soviet Union mentality, that cold war absence of options, of freedom, in short. Not entirely condemned – we go to many international conferences, and host many – but we are always ultimately casting wary eyes over our shoulders."

Chapter 69
Despair in the Afternoon

However unlikely it was, this one survivor of survivors was that of a man – Simon Lev – whose life equated with nothing less than a miraculous property, or properties. The survival part, Taman sensed, had been figured out over excruciating years. It was the combination of the mushrooms he was eating and the invertebrates he consumed, either dead or alive.

There were at least one hundred beetle species in the deadwood. Taman had observed squirrels coming upon such beetles and eating them immediately before burying their red acorns. The whole science of squirrels and boars and, it turned out, wisent, had embodied in their life histories this perpetuity of life within death, which Simon had also acquired by way of both taste and resolve, of intuition from childhood. For Taman, all that was a certainty.

His fascination, however, was most certainly tainted by all his knowledge of the circumstances that brought Simon to this situation in the first place.

Crawling into massive fallen trees, taking shelter, obtaining his protein the hard way, the ancient way. To learn, over time, how to sleep like a log…

That know-how was what someone, or some corporation, was after. For Taman there was no doubting it. Not the names of the Nazis who perpetrated the mass executions. And anyway, more than likely, such information had been subsumed into decades of legal proceedings already, although there had to be many still at large who were culpable. Dozens, possibly hundreds of individuals. Some of them living within a hundred miles of the forest.

But also survivors in Miami and Haifa – people with numbers still burned into their arms and hearts, their memories and their life histories, who remained trapped in the purgatory of any number of litigations. But what of it? Who but they and their children and grandchildren, cared anymore? The ruminations of one utterly flummoxed Belarusian, who had his own problems right now.

What Taman feared most was a state-owned corporation intent upon finding their living subject. This, he believed, was at the root of the crisis he, Lev, and Simon now faced together.

He feared for the man he'd brought forcibly into this maelstrom. How could David Lev ever forgive a world that was more likely to emphatically deny any sense or feeling about human rights or psychological issues involved in the world according to Simon; vast fortunes vs. *capture myopathy*, the cascade of likely physiological consequences – a cascading terror – that would surely result from the sheer shock to Simon's system, being hauled out of his 75-year-long lifestyle, like an elephant chased, hunted, snared with chains and dragged for days and days away from a farm, which he had read about in Southern India, where the Asian elephant was becoming extinct, or, if not killed by the wounds, then shipped to a zoo, to be alone and die slowly decade after decade. Over a million known wild animals languished in more than 10,000 zoos worldwide, the ones recorded in professional associations.

Taman wrote out in his mind's eye each scenario, and this corporate zeal exceeded by a long shot anybody's quest to get their hands on his great-aunt's diary, with its incriminating names that might draw no more than a yawn in the twenty-first century. Most had died, after all.

Moreover, as he had thought about it in the days following his discovery and sequestration of Sarah's little booklet, meditating on its astonishing contents, he came to the realization that none of those mentioned could be tied to any conceivable convictions in a court of law – not in Poland, not in Belarus. Sarah was dead. The names were of children, not provable, not convertible to fact, or conjecture.

For some time he wavered over this crucial set of implacable discrepancies of logic and time, implications growing deeper in his gut.

"What's wrong?" Sofia had asked him one afternoon, when her husband seemed lost, despairing, too many days and nights to remember. The tally of such mournfulness had grown by too much recognizance to be easily written off. His manner had become the weight of some massive disaster.

He had become addicted to sleeping pills and antidepressants long before Rio. His mother, Anna, had little tolerance for his outbursts and vacillating mood swings on the nightly phone calls. You need help, she'd told him. Let me speak with Sofia.

"What did she say?"

Sofia, chiming in and adding to Taman's mental disarray, conveyed the gist of her mother-in-law's scolding: "Pardon my German," she said, "but those fucking mushroooms are getting to his head!"

Indeed, given what was out there – the toxic infrastructure whose syncretistic build-up of spores in any given windy afternoon from as yet unknown fungal interactions could literally cause an autonomic shut-down of one's breathing musculature, causing a heart attack, or, at the very least, severe grippe, or potentially fatal bacterial strep; they had built up a pattern of fear in him.

Sofia had noticed it easily enough. It came in tsunamis of eruptive emotion and he had to hide himself away in his study, lock the door. Sit on the toilet. Vomit in fear. Then go for a walk on a freezing night, just to shake the metabolic heat out of his system that was causing his entire former self to go AWOL.

He was certain that they – someone – would easily kill to get at it, except that they could not find it without him, even with all the available technologies now for surveying forests with GIS tools. They needed Taman, who needed Lev. And Rio was the moment to grab them. Except it didn't quite work out that way.

Chapter 70
Wise Men

Taman knew that without the living being, they could not sample him. So it was up to Lev to find him. And it was a moral duty on Taman's part that that never happen.

This conundrum embodied numerous and inherent contradictions. Nothing made sense anymore. Of course they would find him. They just couldn't sample him. That paradigm was long gone.

He had rehearsed all this in his mind. Until the morality part disintegrated. That's when he truly broke. And Sofia told him that a brother's keeper was not simply some phrase from the *Evangeliary* as she remembered the many tedious homilies and apologies from her Church-going childhood; Latin versions of the "Book of Genesis," and all that talk of the first man to be murdered in the world. Of a brother who killed his own sibling, then lied to God about it.

The Holocaust had enshrined the fiction that was Cain and Abel at so overwhelming a level – the level of Nazi Germany and everything that could possibly be theorized to account for the unaccountable – that she felt it her own insane responsibility to get Taman to Rio, to force this thing upon their family, get it out in the open and over with, even if it meant the end of his career, and the end of their lives – which was a distinct possibility.

Sarah and her diary had literally driven Sofia's husband to absolute despair. And God knows what it was likely to do to Professor David Lev. She and Taman had lived in total terror, never venting a word to their two boys.

There was no way she could ever condone her own husband's withholding this bombshell. Forget the patent, the precarious, ever-so-distant notion of any commercial exploitation whatsoever. It was the man – out there, all those years; a survivor. One could not walk away from such a man. It was unthinkable. She, Sofia, would not permit it.

Her husband, the man she loved, was of the identical persuasion. He needed no coaching to that effect. Indeed, this survivor might well be within six degrees of his own family's traumatic truths, as they say. Which meant that the professor himself,

David X. Lev – he had learned of the "X" by the most labyrinthine of ways – was possibly, distantly related, although this was never actually explored.

And so, with savings – because the Academy could not, or would not finance the trip – he had paid his own way to the Rio Summit. At least it was not time off; they gave him that, and made him an official member of the delegation, but, in retrospect – following the assassination attempts – it had all been planned. He had walked right into a trap. None had anticipated the subsequent events.

Lev was staring at him. "Taman?"

And Taman tried to get past this internal outpouring and reconstruct his original thought, "Where was I?"

"You were talking about the European bison. And the beer."

They remained silent together, their relational psyches that had embarked on a silent warpath of the unknown.

"Beer, no. I was not really talking about beer, David."

"There are huge risks here for all involved," Lev began.

"It's not simple. Your brother is living with bison. The wisent. Strange, but it is the truth. I am so sorry. Sorry about everything."

"No. Absolutely fascinating. Amazing, really."

"Yes. It is."

Then, "So tell me what is really going on out there. Tell me about the wisent. I was a child. At best, I saw maybe a few from a distance."

"They were probably extinct in the wild when you were growing up. It would take years to re-breed them from one, actually two genetic forebears. Their lives? Well, what can I say? It's not my field, of course. But I have watched them for days, months at a time, season after season. They are truly beautiful. Not beasts, but, I like to think, wise men. Wisent. Wise men. That is how I think of them."

"I like that."

"The males, you see. They have already gone, they go it alone for four, five years, like moose – the females and their calves graze upon it. Is there alcohol? No, no. But maybe, who knows, maybe, yes, perhaps it keeps them happy. A kind of bush-grass, good for both grazers and browsers, shrub-like in protein, not that other ungulates populate Belovezhskaya, except the odd goat and sheep on the fringes, in farms where there is still the La Dehesa tradition, of sorts. I have seen twenty-five of the wisent feeding on it at a time, along with sedges, slime molds, liverworts."

"Interesting. Beer."

Taman smiled. "Actually, something like that. No doubt. They are understandably addicted to it and it grows in great abundance here. That is how, I am sure, on a September in 1752, Augustus the Third looked on while his soldiers slaughtered 42 wisent at one go, over 45,000 kilograms of fresh meat in nine hours. Who could eat 45,000 kilos of wisent meat? But there is no surprise in any of it. Russia invaded Poland in 1795. I am constantly coming upon newly discovered Kurhany throughout the forest."

Chapter 71
The Graves

"*Kurhany*? I know that word!"

"Graves."

"Graves."

"Mass graves, usually; sometimes solitary," Taman replied. "A kind of *desertum*, which the forest means, at least in Latin. Empty. It is not empty. Everywhere, corpses. Living repositories for successive maggot and fly and coleoptera species. Carabid beetle infestations. Transit stations for all the pheromones and microbes, and, of course, the bones, they remain."

"Graves. Yes. I remember."

"Of course, it is anything but empty. Sometimes there are human skulls beside wisent skulls. Never in the history of Eastern Europe did pistol-swards – hand-held volley guns in the service of monsters – turn over, give way to ploughshares. Brilliant Polish poets, novelists from St. Petersburg, violinists from Budapest, priests and Rebbes of Warsaw, wisents, Europe's Brown Bear, wolves and lynxes here in Belovezhskaya Puszcza, all slaughtered together. The trains that carried songs, *Yiddishkeit*, precious *kiddush* cups passed down from generation to generation, eccentric recipes, brilliant Jewish traditions, perpetually polished samovars."

"I know. Sasha and I have one. Her father left it to us."

"Well, those same trains carrying that beautiful intelligent, elegant world of three million Polish Jews to the death camps also took hundreds of thousands of dead animal corpses from this forest. I don't know what to say. The analogies are potentially inflammatory but the truth is not: the Nazi ideology was to kill 'The Other'."

"There is nothing to say. There is everything to say. It's insane. There, I said it. The words hang there like court jesters. The sum total of the Fool. Backfiring. You go on. You can't go on. Samuel Beckett, of course, phrased it more prophetically and precisely – differently, but in the same direction, as I recall. I mean, is this not the requisite road to damnation, not just for the descendants of the criminals, but the incrimination of our entire species? We're doomed. I've been criticized most of my life for suggesting it. Here I am, staring it in the face: I am in the middle of it. That's how it feels. A kind of self-perpetuating bathos. 'Get over it,' I would like to tell myself. But I can't. That's never going to be a possibility."

Lev seized his pack, removed a plastic bag, fumbled to open a small container, grabbed his water bottle and swallowed some pills as if in a wrestling match with his own body – the medicines that had always helped to tame his fits of anxiety that resembled heart attacks if he did nothing.

"Just give me a minute." He removed from a zipper pocket a bottle of meds and placed the nitroglycerin tablet under his tongue.

"You all right?"

"Not at all."

"Angina?"

"Everything. I'm on several vasodilators."

"The drugs help?"

"Hope so. I also wear the patch – two, in fact."

After some time had passed, Taman offered to change the conversation. He wasn't sure what to do, fearful that Lev could collapse. He had considered the ramifications of an 87-year-old coming to the forest in winter, under these...unprecedented circumstances. All he needed was a dead, famous scientist on his hands. A colossal misfortune. Then he shook the demon away.

"Today, Poland celebrates Jewish tradition, to varying degrees. Yeah, really. Several years ago Prince Charles himself – I forget who told me – opened the Jewish Community Center in Krakow. Now, every year thousands of people come to the Jewish Culture Festival there. And it's no different in Belarus. Parts of Lithuania, Estonia, even Russia. Everybody wants to prove how cool the Jews are. I think it's good for the economies. Like eco-tourism or something, these bouts of the odd Renaissance of Jewry. I read an article in the New York Times about it."

"So how many Jews are actually left in Belarus?" David asked.

"That's a good question. I would think no more than twenty-five thousand."

"I looked it up. Most who could manage it got out, to Israel. Made haste."

"We have lots of Jewish friends in Minsk. But, really, I just don't know how many are left."

"Do they know you're Jewish – part Jewish, I mean?"

"Of course. As a scientist, however..."

"What?"

"Being Jewish doesn't give me bragging rights."

"You're losing me."

"Maybe they don't know about it. It's not like a subject of conversation. And I'm in the field so much. And maybe it's still worrisome to openly refer to it. I don't know, David. Since this business with your brother,"

"This *business*?"

"Please. Stop being so judgmental. Defensive. You know exactly what I'm talking about."

They sat silently. After some time, David mused discordantly, "She loved my little brother. But I cannot remember her?"

"Here." Taman took a picture of his great-aunt out of his wallet.

David held it in his hand. But he shook his head. "Amazing. I have no recollection."

"How could you? But clearly, she and your brother were very close."

"All children are close. It is a universal unity. Nobody can take that away. Not even Darwin or Aristotle would have ever thought to try. Quite the contrary."

The large metal door swung open and Ulyana strode in, slamming the door behind her, trailed by a loud roar of the wind, and the churned-up snow – as of silken embroidery on the fringes of a great bridal gown – racing under and through every available chink in the threadbare nature of a human body, all vulnerable and exposed to the least challenge by nature.

"Ja dumaju, što bura prachodzić. Ja daju jamu tolki niekalki hadzin."

"What?" Lev queried.

"She thinks – and I would agree – the storm's moving south/southwest and the worst has passed. It's going to penetrate the Polish side in a matter of hours. If you are up to it, we should get going first thing tomorrow."

"Twelve foot snow drifts, isn't that what she said earlier?"

"Yes. I apologize for our weather." And he smiled, maybe for the first time ever.

Chapter 72
52°35′22.86″N 23°54′12.54″E. Elv. 201 M.

Taman knew the coordinates by heart. He didn't need the help of any electronic or battery-powered device. Not after five years, and certainly not now.

Lev knew that only Simon could find the trench in particular, the one that held their parents, grandparents, cousins, aunts, uncles, the local Rebbe, and Lev's own girlfriend from back then.

But only Taman Chernichevsky had a chance in locating Simon Lev. The weather was absurd, thought the professor, absolutely out of a Hans Christian Andersen fairy tale, but with no princess, no Thumbelina or Snow Queen; it was Peter Pan without Tinker Bell, the least glimmer of hope, of flight, of anything but pain.

The landscape was, in a word, desolate, beyond any word in any language. Lev was certain Mallory's Second Step upon the northeast of Everest must be easier than this. As for Hillary's Step, near the summit, now, he had heard, an artificial ladder had been secured to the rock wall for all the tourists.

For eighteen hours his pulse had pounded at well over 150 as he worked to keep up with his much younger ally on ill-fitting snowshoes amid 1,250 square kilometers of frozen forest, an area not quite half the size of Yosemite National Park but vast enough.

They were within three kilometers of the giant pedunculate oaks, a few standing, others dead, Taman had explained carefully. This was the turf, he said, where Simon was likely to be.

"If the weather had permitted, we would have come sooner. I have done my best."

For Lev, the landscape had long been subsumed in his mixed memories as if no time had elapsed: flashes, he was there, back then; in living, breathing encapsulations of every sensate truth, as they erupted mystically in his gut. Ideas are fixed; they don't age. Senses don't age. Lev smelled, remembered, touched every identical element in the forest.

Even as he struggled through the challenging wilderness of icicles, frozen bogs, hanging panoplies of otherworldly shrubs, berries, and leaves white and arthropod-like, in their web of hurricane-speed paroxysm, as if the paranormal had frozen to the air currents, he remembered this living relic that yielded to myriad prehistoric curtains of hundred- foot-high frost, frozen air that enshrouded this mist-enraptured wonderland of brutalities and haunted marvels.

"You remember it, don't you?" Taman said, labored breathing, frost elliptically wafting outward from each word inches like philosophical cigar smoke before his chapped lips as they uttered the meaning that could not be more clearly evoked in the measure of Lev's deeply solemn sense of place.

The temperature was nine degrees. But to Lev, working as hard as he was to keep up, it was damned near balmy.

"I had forgotten. Now I remember." He tied his parka around his waist.

"What are you doing?"

"I'm roasting."

"You nailed it."

"I'm not a mind reader, Taman."

"That's the secret to all this."

Chapter 73
Eyes on the Ground

Jake and Claire, with Allan Hobbes and the DD standing behind them watching, manipulated the new app from inside Jake's large and elaborately outfitted sanctuary at the Agency.

"We're eyes on the ground," he exclaimed in a kind of rush he had not experienced since first learning "Netscape Navigator 2.0," a version of Javascript, some twenty years before, or, much more recently, mastering all of the cookies inherent to Landsat 8 Operational Land Imager (OLI).

"How are we hearing this?" Claire marveled.

Neither Hobbes nor Norwich had a clue.

Jake summoned all his tricks of the trade, manipulating with a free hand to deftly augment what was already miraculous. "I've never held this kind of joystick in my hand before," he confessed. "Better than a Ferrari."

And what they saw, and heard, were Taman and Lev making their utterly and inexplicably difficult way through the blizzard and – with an ease of zooming out, the sight of two individuals, unconnected, at distances of approximately 2,000 feet, in the case of a woman, and 1,200 feet in the case of a man, each well armed – apparently following the two scientists, the followers clearly unaware of each other.

"We're seeing them through the shadows!" Claire exclaimed, totally astonished by their newfound toy.

Jake slid his finger on one of many pads on a mainframe that came with the new app. "Listen!"

"My God!" Hobbes was heard to say. "I can hear their actual breathing, separate from the storm."

Chapter 74
Plants That Dream at Night

Taman had been talking to Lev about so many of the plants, masked in winter, but known to him nonetheless, in part, because many maintained high metabolic energy levels even deep under the snow: the many carnation, buttercup, leguminous and sedge family composites.

"Sometimes, you'll think I'm a bit out of my mind, but I hear them, as if they are dreaming – and at night, if you listen, you can hear them."

"Yes. You are definitely mad," Lev said, utterly out of breath. "But I can respect that."

The forest had never left Lev's mind from summery youth, in full bloom – the birch, the black alder and Norway spruce. In lime and elm trees, a kind of sanctified primeval painting that mirrored the canvas of universal childhood, many acres of the divine. All driven by variable energy needs and expenditures, irrespective of the ambient temperature.

"He has found his own methods of staying warm. Just like the wise men. He removed his parka, I suspect, within weeks of having to survive."

"You're referring to Simon, I presume?"

"Of course."

All those things Sarah had written at one point in her diary. Sarah, whose face appeared to Lev finally, as he stood beneath a vastly stretching oak (Eastern European, not French, not English), not one of the huge trees named after some Tsar or other, but just one of many hundreds that were six, seven hundred years old.

He thought he could finally see her. A vague little smile, tentative, at best. But, even then, somehow flirtatious, as only a little girl, possessed by the wondrous demons of a forest like this, can evince.

But in recalling that little girl who had become an old woman in a nursing home, it reminded him, covered in frozen hanging drifts and vast panoplies of icicle, of the General Sherman Tree in Sequoia National Park, all 1,500 cubic meters worth, one winter – winter by Sierra standards in the old days – prior to perpetual drought. A breeze, in other words. And of a picture some tourist had kindly taken of him and Sasha. The way she leaned on his shoulder under that tree, blushing, even after all those years of marriage.

And of the Black Wood of Rannoch, in Scotland, where Lev had visited the last stand of Caledonian pines one icy winter afternoon; or of Sherwood Forest, what was left of those ancient oaks and pasture surrounding a village in the north-central U.K. called Edwinstowe, also visited in winter; the few remaining groves of the Cedars of Lebanon, and others of their rarified ilk that Lev had had the rare privilege of seeing throughout the world.

None of these ecological contexts meant a damn, at this point. Not for an 87-year-old in search of his brother, without a particularly well-thought-out reason for going after him.

Going after him…

That's what he was doing. What they were doing. By all rights, it was a horrible thing to be engaged in. To what end? Why do it? He stopped solid. Not a single inch.

"Taman. What are we doing?"

Taman became motionless, leaning on his predominant ski pole catching his breath, turning back to look at the professor. "What do you mean?"

"I mean exactly what it sounds like. Is this the right thing to be doing?"

"If you are truly asking me an ethical question, I must answer you, but you must also know that I am unaware of any situation comparable to this, ever. So whatever I say is bound to be wrong, at least in somebody's estimation. Probably in yours. David, I don't know. I brought you here. You agreed to come. For that, how many times must I apologize, in how many ways? And what about you? We're here. There is no point belaboring this."

"I'm just not sure, Taman."

"Then let me at least suggest the following."

"I'm listening," Lev sighed.

"Your brother has transcended all of this shit. What I have discerned – discerned, that is the word, in English, I believe?"

"Uh-huh?"

"He, if anybody among our kind has ever accomplished it, he has done so."

"Accomplished *what*, Taman?"

"These plants that I have studied my entire life – your brother, he lives with them. They dream at night. The wise men, the wisent, let us be clear – European bison – Paleolithic bison – they all dream together. This is something we no longer understand. We have lost access to it. But your brother, by all these truly evil events, he has penetrated into the core, and he has, for certain, become one with them. It is a miracle, David. And no. It cannot be bought at a pharmacy's counter."

Chapter 75
Mystiques Amid the Rhizosphere

It was their fourth day/night in Bialowieza. Lev could not keep track of the passing hours, not with their getting up at 3:25 in the mornings, and with continuing snowfall that blurred most distinctions between light and dark. A continuous blasting haze of whiteness.

In the midst of the worst winter on record, the countless great trees were festooned in the thousands of bryophytes, epiphytes, club mosses and ferns, the fungi and mystery hybrids that Taman had devoted his life to.

Lev had never seen such a forest, at least not through professional eyes. Little wonder that prehistory was writ so large throughout the tumultuous history of this primeval world, the last in Europe, dating probably to those most ancient of seed-bearing plants from Ireland's Middle Silurian period, and the group of 385 million-year-old *Eospermatoperis*, stumps revealing genetic evidence for the earliest forest ecosystems in the world, the cladoxylopsids of which Bialowieza was the sole European descendant, and through which the two men now labored before plopping their butts down to rest in the deep drifts.

"Here, let me show you something."

Taman sat beside a rotting old tree. He brushed the low angle of drift away, revealing an actual melting line, an upper cusp in the icy tumult.

"We know of at least 162 aerophytic algae, the ones inside this kind of bark, and an unprecedented Cambrian-like explosion of fungi which have been somewhat key to my genetic work out here."

"And how does it work? Really? Well, give me the one-minute version; I'm fucking exhausted. But I'm just sitting here, so you have a totally captive audience. At least until I'm ready to stand up again."

"I've put instruments inside these trees – not this one, but similar – in order to obtain electron microscopy of the insides of the deadwood, much like in a hospital combining PET, MRI, and CT. It has enabled us to estimate the amount of CWD, course wood debris – about 25% usually – and what life forms are inside."

"So what's new about it?"

"Well, I now believe that the algae, the beetles, and the microbes account for some 80% or more of the life in this forest. That, my friend constitutes an enormous secret, a discovery that accounts for your brother's survival. Welcome to the world of the Late Devonian," Taman went on, as a frozen evening settled upon them. "You are now a bona fide member of a unique place: you are inside a commune of life, but which most people, particularly the forest managers across Europe, consider to be dead. You are now in the kingdom of literally trillions of individuals known as saproxylic life forms. They don't bite. The larvae of Britain's famed stag beetle is one example. Your brother is eating such larvae. He's even adding delicious spices, I would imagine."

"Such as?"

"Garlics. The Polish wood sorrel seeds squashed and mixed with a little oil of camphor, *Oxalis acetosella*. And the pollen of forest apple, small-leaved cranberry and gooseberry, for starters. I can't even imagine the new species of macrofungi he has probably experimented with, after 75 years. The soft, delicious sporocarps – grant you, a developed taste, but one heightened by the mineral salts in fungi, not to mention a million edible spores from mushrooms yet to be discovered. The boletus, obviously – well known and loved. But also the honey mushroom and calvarias. Brilliant morels that will set you back thirty dollars for a bowl of morel soup at the best restaurants in Belarus. Especially now, with our new administration, budget cuts, and taxes following upon the crash of the ruble."

"Surely we were not targeted in that café by lunatics trying to figure out what my brother is eating out here?"

"I confess, I'm not sure. At first, I really thought it was the naming of names in that diary. But I have, you see, a patent pending and I don't really own it because of the nature of the Academy of Sciences. They own it. My wife and I think it terribly unfair. But, in truth, they own us, and my two sons."

"What kind of patent?"

"It's related to root exudation. The strange world of the rhizosphere."

"How strange are we talking?"

"Strange enough that it might well be a template for improving human memory. Speaking of which, do you remember what I mentioned about the bush-grass and the wisents?"

"Which?"

"Your brother, David, appears to have grown up, been surrounded his whole life, by these unique and gentle giants. Yes. I believe he is a true wild child, possibly the only one in known human history to have been raised by wisents, the European bison."

"I still don't understand the patent you're talking about? The wisent live, what, 15, 20 years out here?"

"Yes, but how old is your brother? And by what seemingly miraculous survival stratagem has a child survived 75 winters like this one? It's going to be well below zero tonight. And he is out here."

"We need to find him."

"We will, if we can. Let me answer your question. It's all in the root exudation."

"Yeah, you said that. Oozing organisms. Pretend I'm an idiot. Explain."

"The roots are speaking to insects, to other roots, to parasites, fungi, beetles, to bark and to soil. The world is communicating under our feet."

"Ever ecologist knows that, sort of."

"Do you really know? Because I have tried to measure it and I have no doubt, none whatsoever, that it is happening and your brother has tapped into the whole realm I'm talking about as sure as he has tasted the taproots that are part of a system only now beginning to be grasped by the scientific and agricultural communities."

"You're saying Simon is speaking to roots, fungi, to beetles, and to the frozen soil?"

"Yes, I mean, sort of. It's…it's emotional."

"He is a Holocaust survivor. Anything is possible. But just to be clear: Everything we hope for is ultimately impossible."

"You believe that?" Taman asked, taken aback.

"Don't you?"

***

"This does not feel right," Claire said. "This is what happened to us, Jake, in our own kid's bathroom."

They were alone in the new "app opps room," as they were calling it, five stories below the most highly secured road in North America leading from the George Washington Memorial Parkway.

"This is different. You, we need to get over that."

"I need to see something."

"Where are you going?"

"My office. I'll buzz you."

Minutes later, Claire called Jake on his newly installed app opps protocol site. "There are at least half a dozen Belarusian military vehicles amassing at the gates."

She was watching on two mainframes, frequencies, like huge EKGs.

"At the entrance to the park?"

"Yes. But I have that from chatter, not visuals. I'll be there pronto."

***

"Look," said Taman, exhausted, reflecting on a very bad and blurry number of previous days, weeks, all leading to this dreadful nameless vortex of ice and misery. "I can't follow up on such a feeling. You know that. No one who is responsible, alive, conscious, in this century, can answer to that. Our collective is a disaster. Individuals, okay – some of us at least aspire, we try to redeem the concept. And maybe it is possible. I have to believe it is possible."

"The concept of what? The possibility of what?" Lev managed, out of breath, coughing.

"Of us human beings. Doing something more than being born, killing one other, being cruel, and then dying."

"Very unlikely, scientifically speaking."

"That is a horrible vision of evolution."

"I know. Maybe it comes with age."

"No. I had a teacher ten years older than you. He survived the Holocaust and remained an optimist to his dying day. This astounding acuity – that mycologist whom I treasured, who taught me so much; your own brother who is clearly astounding beyond all logic or any possible comparisons that come to mind – such men must be understood, proved, exclaimed to the world. It is what science, certainly biology, has always strived to discover. Plato, Erasmus, Darwin, Humboldt, Buffon. It is the holy grail of ecology, as you know. Invertebrate and plant species speaking to each other. And to us. Come on! Biological molecules articulating themselves into the conversation? It's amazing. Fuck Hitler."

"You're quoting Churchill's parrot, I believe. But don't confuse the worst moment in the annals of biology with … with any of this."

"But their corresponding moment in the history of life cannot be mere coincidence."

And Taman went on to describe his simplified story of red squirrels who know exactly where their thousands of acorn caches are planted. David Lev was all ears. He had nothing better to do, other than try and keep any more heat from escaping from his badly tattered body.

***

"Turn off whatever is recording their conversation," Claire insisted, seated beside her husband in the app opps room.

"I can't"

"Jake, turn it off! This is inappropriate."

"There's no override, all right? It captures everything. That's the whole point. And it might just save their lives."

Chapter 76
The Genius of a Squirrel

The snow falling upon their heads, their spirits, every inch of Bialowieza, would not cease. These hallowed grounds had turned to miserable doldrums in whose eerily silent Epidaurus-like expanse, Taman expatiated at length on a topic dear to his heart, if remote from their present crisis.

And it was a crisis.

Within the hairlines of sight, two people watched their every move: One from the south, another from the northwest. Petrovsky was positioned in a bivouac that enabled him a perfect line of sight. He carried his crossbow with the stealth of an Olympic archer.

Ulyana, who had been following the two scientists as well, kept two rifles at the ready, a third in her pack.

Both of them were so well camouflaged as to elude each other – quite a feat, considering.

Thirteen kilometers behind the two scientists, as viewed from space, and from the app opps room, 14 Belarusian military, accompanied by John Vespers and his Chinese counterpart, no dogs, per se, were now headed on foot from the group's abandoned convoy of Typhoons and Maz-543 Uragans whose Arctic-grade maneuverability – vertical steps, engine power, trenches, etc., could not withstand over a dozen feet of wet snow. Their goal was to surround the research station. They did not intend to follow the scientists.

Vespers – and only Vespers – knew his girl was out there, and more than qualified to deal with what was inevitable, according to his own real-time intel.

***

"I've always wondered about the co-evolution of the squirrel brain and condensed tannins," Taman mused, nibbling on a protein bar, which he then offered to Lev.

"More importantly, there is something happening between the wisent and the squirrels, the boars and the jays."

"Taman, I'm tired."

"Just making small talk."

"Whatever."

"So, you see, these scatter/hoarders" – which is when Lev recognized all the signs of a beautiful sort of craziness-slash-genius in his hyperactive and exhausting companion – "they know where the white versus red acorns are hidden; they understand precisely how to dry out the mushrooms in the trees. The bison are grazers, as you know. But even this time of year, when it's so bloody impossible, they must dig and dig they do – it is painful to see – through this deep snow, desperate for any sedge, grasses, truffles."

"We haven't seen them?"

"We will. You made me lose my ... oh, you see, here is the magnificent point, David. There's something new in the biomolecules that is attaching to the proteins and amplifying their mental maps. They talk about it."

"Who, Taman, you're losing me? The squirrels or the bison?"

"All of them. I believe all of the vertebrate herbivores."

Lev was as fascinated by the content as by the frost in the air, Taman's breath yielding to sub-zero cold. Indeed, the hard-breathing professor was inebriated, the first pangs of hypothermia. He actually didn't feel bad, just slightly drunk with the cold. Feeling a sting that could be interpreted as either cold or hot.

Lev also knew deep in his own data bins that is was precisely the vertebrate herbivores of the world who were most rapidly verging on extinction.

Then delayed in all his actions, response time, Lev inquired, "Talk about what?"

"David, try to imagine the following scenario: He, your brother, he understands what they're saying. Where the food is. Follow the food, we find your brother."

"But I thought you said up in a tree, a particular tree?"

"He has moved around. Of this I am certain."

"Then how the fuck do you know where we're going?"

"I know. Trust me. We're very close. And as for this cold, he has found a way, not by growing fur, obviously, or even living wrapped in the hides of deceased bison. But he appears to have fashioned his own bodily costume to ward off the weather. I believe I know approximately where he also maintains the equivalent of an igloo, inside a huge dead tree cavity lying on the edge of a frozen bog. Cloaked in what must be the equivalent of seal-skin for Eskimos. That's where we're headed."

***

Jake's mouth was agape as he and Claire watched and listened from app opps.

Jake had learned the essentials of this new technology with lightning speed. What was most impressive, and frightening, was the sound-wave consolidation from millions of digital files; amplification, flattening of the acoustic waves with some remarkable equations capable of lining up picture and sound from, literally, the head of a pin in the earth's infinite hayricks.

"Adiabatic decompression," he explained.

"Lost me, kiddo."

"The longitudinal waves."
"I don't have NASA on my CV, remember."
"Anyway, this is fucking crazy."
"It's not right," she continued to proffer.
"Yeah, it is," Allan Hobbes said. He had just come into the room to get an update. "Enjoying the show?" he asked.

***

"All entirely logical," Taman went on.
"You had mentioned Edward Curtis. Various Eskimo tribes?" Lev replied.
"Precisely. In this instance, decaying matter from the deadwood makes for a metabolism, the breakdown in biochemical terms of nutrients – if you are a squirrel. A human, this is a first in Bialowieza, at least in many centuries. You see, I don't think he has ever tried fire. It would cost him being found out, routed, amid a vast complex of roads, a train a mere 35 kilometers away that carries chemicals, a border crossing, a reservoir, park stations, tourist routes. And a shitload of military redundancy."
"What are they so worried about? Becoming the next Ukraine?"
"I don't know. It's complicated."
"Everything's complicated."
"Especially your brother. David, I am quite certain he has learned the secrets of deadwood; of silence and of stillness; of that majority of warm exuding life forms that call tree hollows home. Nobody ever thinks to look inside the huge tree hollows. Here most people are frightened of them."

No fire. The concept found in Lev a rallying cry, for he had obsessively studied it in and out of his many navigations, throughout an endless career. From the highlands of New Guinea to Arunachal Pradesh. From schizophrenia and fire studies at the Johns Hopkins Medical School, to the annals of a certain schooner in the Andaman Sea, back in the 1840s, where a tribe without fire was most definitely discovered.

To imagine that his very own little brother had never had fire ... it defied, not just logic, but, well, the human body. Yet, there it was. Taman was certain of it. And the circumstances seemed to comport with his reasoning: Had Simon ever lit a fire, he would have been found out, murdered, like everyone else.

Then there was the issue of the tannins. A co-evolutionary function involving tannins.

"Explain?" Lev said weakly, blowing into his cupped, gloved hands, for even a modicum of passing warmth.

"The condensation of flavans. Polymers. Sorry. It's something of a pet project. The truffles may be crucial to rodent and mammalian memory out here."

"Uh-huh."

"I've even suspected anterograde amnesia in the old boars. You understand, I think?"

"I don't understand a fucking thing you just said," Lev barked. For the first time he began to really appreciate the idiot savant-like crazed mycologist, who was not unakin to his own polymath-like umbrella of interests, obsessions and expeditions in younger days.

"It's how I happened upon him, once, twice, three times. And how you'll be reunited. Follow the wisent."

"Reunited? Are you out of your mind?"

"You're here to rescue your brother, are you not?"

"Taman, put your own gun to my head. To his head. He's not a lab rat. And, if my memory serves me, he ate kosher chicken soup as a child, and whatever meat could be had, in our meager circumstances. All your theories out the window."

"But whatever he ate as an infant, more than seventy-five years has surely altered all of those proclivities, you surely understand"

"I am not here to rescue him. I can't save myself, let alone my younger brother. What would that even mean?"

"I don't know."

"Well you better fucking come up with something better than a rescue. Because this is not that and we are not the marines."

Taman sat silently. He wished no further escalation of his own deeply flawed intrusiveness. He just couldn't let it go. And didn't know what to do, how to make amends, how to solve their – his – personal time bomb of a dilemma.

"We'll just go with it," Lev finally said, certainly not breaking any ice.

"Shhh!" Taman put his finger to his lips.

Lev eyed the direction Taman had indicated.

Then they both heard it. A helicopter, flying low.

"In there."

Both men ducked for cover within a snowdrift that provided cover between two enormous trunks of hawthorn.

The noise of the chopper went in and out of the storm.

Lev whispered to him: "You said you were in the military once?"

"Out of high school. Two years of service. Shhh!"

He was listening intently now.

The chopper's roar got closer, then quickly vanished.

"Don't move," Taman said.

"I wasn't planning to."

For five minutes they squatted, leaning motionlessly.

Finally, "I think they're gone."

"They?"

"Your guess is as good as mine."

"What kind of a chopper was it, do you know?"

"Air force."

***

"Guys…" Jason had zoomed in on the chopper.

"It's the Belarusian Mi-17, the standard NATO-type multimission chopper. It'll carry a lot of soldiers."

"They can't put down."

"They won't. It's a scare tactic. Can you spin the cockpit around and zoom in?"

"Let's try."

Within seconds the new app had proved miraculous beyond anything Jake had even considered.

"Pilot, co-pilot, and – two other guys. Ones Chinese. The other…Wait a minute!"

Hobbes buzzed Norwich, but she was in a meeting with the Boss.

"Jake, freeze and store those two images. Claire, please fire them over to our friend Matt at NSA. Pretty sure I've met that guy."

Within minutes they had confirmation. "That person is John Vespers," Hobbes asserted.

"He's our Henry," Jake declared, throwing a glance at Claire.

***

The two professors continued towards their destination, moving through gigantic obstacles. Taman was deeply impressed by Lev's pertinacity. He also noticed that Lev had downed probably half a dozen of his pills, maybe more. He sensed the worst, knowing what lay in store. But these were ineluctables, and if indeed that were the case, it was now far beyond his grasp to change the basics. Meaning, an 87-year-old dropping dead in a wilderness, concealed by the worst winter on record.

Taman tried, however awkwardly, to make some kind of near religious mends, through nothing more than the articulation of a certain silence, gestures of complete fellowship. Even the occasional pat on Lev's back.

Then, as they reached a resting spot, he began, "They raised him. The bison. I am certain of it."

Chapter 77
The Wisent of Bialowieza

Both men focused on the sky. Taman knew they were being tracked, but that it would be difficult under such difficult weather conditions. The forest was thick, labyrinthine, and in these conditions, and in this portion of the wilderness, there was probably no one but Taman – and Lev's brother – who knew it as well.

"We're close," the mycologist said, out of breath. "The wisent." He was no mammologist, but he knew them. *If you know what they eat, you know their soul*, he'd once expatiated before a classroom of undergrads in Minsk, all of whom, of course, had grown up with the mantra "You are what you eat."

In this case, a great bison whose life histories had come to dominate – for so many centuries, indeed, hundreds of millennia, in and out of ice ages – this last relic forest in Eastern Europe, was, when everything was stacked up and, if ever to be judged at some pearly gate, the gentlest of all creatures in the forest. As if, were there a God, the very *theory* of gentleness had been tested in one of the largest terrestrial creatures on earth: Did evolution support non-violence, timidity, enormous biomass wrapped in sentient tenderness? The wisent proved that theory.

Gorgeous and elusive, grazing atop the Lost World, confined innocently to their own Pleistocene Island, they should, by all accounts, have ruled, Taman always reckoned.

Taman openly reflected, between the haze of gelid air, frost rising like a fog surrounding the two laboring men, "They move back and forth, several kilometers each day, sometimes more, depending on the season. There's probably a third more here on the Belarus side than the Polish side. Last count, about 564 individuals. They could easily become extinct, at least in Poland. Prior to World War II, as I told you, they almost did."

"And worldwide?"

"A certain fascination with wisent has grown enormously, as you know, David. My guess is there are probably 2,800 in the wild, well over 4,800 if you include captive wisent. The Polish bison – and, by the way, we are now just inside the Polish side if my senses have not betrayed me – are descended from seven individuals."

"We're standing in Poland?" Lev was stunned by the sudden realization of how close he was at that moment to the village in which he spent his first eight years.

"Yeah. We're in Poland. So here's the deal: Two Y chromosomes and their genetic inbreeding is increasing. The true żubroń herd, normally, there's no more than about twenty together. Tragically, as trends escalate, it's over for them. I give them ten years."

The statistics just flew off Lev, like dry snowflakes. He was focused on his brother.

Then, regaining Taman's focus, "What? You were saying? What's the limit of their tolerance to cold?"

"Thought you'd never ask. Y chromosome-related, I suspect. But it varies in the two genetic lines that have been ascertained. And also there is variance between the males and females. There has been a certain amount of hybridization, you see. There was a genetic bottleneck. Bison studies in the mid-1970s from your country showed metabolic constancy at even −30 Celsius. Here, I've never seen precise data, but they are related to the American bison, as you probably know, and these animals manage to survive severe winters – this one included, let us pray. In Siberia, somewhere south of Chersky along the Kolyma River in the Sakha Republic, you may have heard about this new Pleistocene Park where they are reintroducing wisent, along with many other primeval species."

"Yeah, I met a research fellow who'd been there."

"Well, they are trying to recreate the Pleistocene Epoch taiga-slash-grasslands to prove, among other things, that hunters, not climate change, drove all these species to extinction. There, the temperatures apparently drop not infrequently to −50°, that's in Celsius."

"And in Bialowieza?"

"It's very cold at the moment, wouldn't you agree? Especially for late March."

"But what about their tolerance? My brother is hanging out with them, Taman."

"Their thick hair; sub-dermal fat content; their fourteen rather than fifteen sets of ribs like the American bison have. Who knows what all that accomplishes in evolutionary terms? But also their group behavior. They go neck-in-neck. Highly nurturing. My theory, based on what little I have to go on, is that they have kept your younger brother warm because they probably think of him as a perpetual runt – an orphan, not an outcast."

"So they protect him?"

"That I have seen. He is brought into their folds as they congregate this time of year. The cumulative insulation of their 1,500 to 2,000 pounds per average body heat, nearly hot breaths, tight confinement, provides the equivalent of armor against this cold."

"A bit like penguins."

"Maybe. I don't know. But then add the compounded option of internal ambient temperatures that rapidly escalate within the tree hollows your brother has been occupying all these years. It has clearly been a successful adaptation."

"It's too incredible."

"Think how you suddenly got the urge to wrap your parka around your waist? Well, he's got to be the world's foremost expert on living in dead wood."

"Are these hunches or actual hypotheses? Is there any empiric evidence for any of this?"

"I have tried, imprecisely, to map the memory areas of the wisent, for they are also followed by the boars, the squirrels, and the jays, each of whom has evolved cerebral maps wherein evolution must surely favor clever memories."

"The birds are going after insects attracted to the hot dung."

"Of course, but it's more than that."

***

"Of course!" Claire exclaimed, rolling her eyes. "These guys are the greatest geeks on the planet! Of course???"

"I have a hunch we're about to see wisent," Jake followed upon his wife's conversion to an app that had, Hobbes let slip earlier, cost the U.S. government collective spy agencies in the billions of dollars to develop.

Collective…That was the megatonnage of a word that stuck with Jake. He realized they might not be the only other geeks on the planet following what was going on.

And then Claire had added her own afterthought: "What's to prevent, say, the Chinese, or Belarusians, for that matter, from hacking this by now no doubt notorious little app?"

Jake had at once divined the algebraic space probabilities, as he was taught at Columbia in an advanced GIS course from a Nobel Laureate whose own modified astronomical algorithms had enabled new geolocators that fixed upon light readings from the many dawn-to-dusk data compression networks of transmitters and satellites. They, in turn, logged all the electrical interference of bird migrations with other objects to recognize distinct human images on the ground that popped out, like drop shadows.

***

"I do remember Simon was something of a whiz kid at cheder. Even the Rebbe commented," said Lev. "It comes back to me, now. At a very early age he was memorizing the necessary lines from the Torah."

"Well here's the thing: If all these mammals – squirrels, wild boars, wisent – have maps to edible white acorns, the embryos already having split open above the moist fertile soils, so does your brother. He has the same map. And, apparently, a fine memory, as you say."

"So, you're saying Simon is eating acorns?"

"No. Yes, I mean some cultures do. Koreans, Native Americans. But the squirrels, mostly. David, my point is, he is wild. Truly wild. Science can learn so much from him."

"That is not why we are here, Taman. We've been over this. I cannot stress my concerns."

"Remember what I told you in Rio about my research? Since discovering your brother, it has forced me into fields I had not anticipated, like zoosemiotics, the cognitive maps of those primary seed dispersers in Bialowieza. Haller's Rule. Brain-body size isometry in wisent. The applications are mind-boggling."

"You've published this stuff?"

"No. No way. I told you. I published all of one research paper on certain mycological systems. But I clearly made one huge mistake: I mentioned it in some detail to one of my grad students, I also told you. He, in turn, cited it, calling it "Personal communication with Professor So-and-So" in his dissertation footnotes. I could not fault him for the citation. The problem is he went and published a crucial summary of his own spin on some of my research – it can happen – but this could have been a deliberate sort of sabotage, in retrospect – or not. I don't want to judge."

"Where did it show up? An established journal or a blog?"

"Unfortunately, not just on the Web, but within the rubrics of a journal devoted to eco-dynamics in European forests. That's how the word got out. He must have known precisely what he was doing, and who would see it. This was no general shotgun spray. This was targeted. It may have compromised my entire research project. Coordinates and all."

"Obviously. The chopper. Everything. If that's what's happened, then so be it," Lev uttered. "Fuck it. We're here."

"So that's what happened in Rio," Lev continued, making sense of it all.

"I don't know. And I don't know if my great-aunt's diary has any connection. Frankly, I doubt it. Nobody knows about it other than me and my wife. But there is that American you mentioned, from the Consulate. In commerce. You said he was weird."

"Very. Although he probably saved my life."

Chapter 78
Nothingness

They moved for many hours. Another day, another night. There were wisent tracks everywhere now.

"Sometimes I make believe that I can understand all the harmonies in this forest," Taman contemplated aloud. "A sweetness, poetry."

Lev could hear him, even though he was some distance, hidden, trying to do "number 2." He hadn't for more than two days. And such constipation, especially at his age, and with his blood pressure, was a serious problem.

But hearing the notion of "sweetness" sat so badly with Lev that it somehow miraculously helped him to accomplish the task at hand, adding, no doubt, to the eco-dynamics of whatever the hell Taman had been talking about earlier. But now Lev was seriously thrown. He mulled over the lasting impression, not a pretty one.

At length, during a five-minute break amid a severe punch of low-pressure theatrics, David finally tried to offer the rebuttal that had been bearing down upon him.

"You believe he sings. Is that what you're telling me?"

"The wisent?"

"I'm referring to Simon. Taman, do you actually believe that there is a Holocaust survivor out there, raised by European bison, surviving against all odds, singing to bugs, bugs singing to him, talking to squirrels, murmuring sweet nothings to the wisents, all the animals in one happy paradise – these fucking snowdrifts – the stomping ground of tyrants, the killing fields for enormous piles – that's right, piles – of Jews shot in the head, mass anonymous burials two, three stories high – is that the song you're speaking of? Is that the great science you're extolling? Some fucking journal by your brain-dead great-aunt implicating my little brother, a victim of unspeakable evil?"

"David, please."

"No, Taman. I have to get this off my chest. None of this smacks of a sweet song. Of fascinating eco-fucking-dynamics. The minds of a squirrel. The biology of wisent. These are not the realities of World War II. Not for Simon. And not for me. Or the memory of our parents and grandparents."

"It is what it is."

"No, it is not."

"But something akin, something magnificent. He survived, David. Your brother is testimony to..."

"To 75 years of terror. Horror. Of hiding out in a nightmare. Nothingness is not science. It is pain. Sheer unrelenting pain. God help him. It's nothingness, Taman."

"I apologize. I'm sorry. I truly am. But..."

"But what can you *possibly* add to this disaster upon disasters?"

Taman hesitated to take it any further. He knew David was absolutely right, on every level. But he had also invested a vast emotional archive of his own in solving what seemed to him to be the most notable riddle of the twenty-first century.

After an hour of trudging silently, the two sat down uncomfortably for more than a few minutes.

Taman appealed to Lev: "Maybe you forget. We all tend to. But for years my research has headed directly towards what my great-aunt announced – without a degree in science. With only her childhood love of a little boy. She simply knew. Your brother clearly knows. And now very ugly people want to sample his blood, his genes. They want to clone him."

"How do you know that?"

"Because if they were simply worried about his naming names, Nazis still thriving in Minsk, or in Paris, or wherever, they would have hunted him down and killed him. Okay? I do not share in that zeal. I want to protect him, because I know what they are after."

"I agree with 'have to help him.' Obviously."

"Then we understand what we're doing here."

"I'll buy that."

"Those same types of people in different uniforms tried to convert Bialowieza into the first Aryan forest. Back-breeding into a vision of eugenic hell. They slaughtered most of the animals in the Warsaw Zoo. A few good people saved as many as they could."

"There were many unheralded Oskar Schindlers."

"Yes. And also a few very good Poles and Belarusians who were part of the many attempts to assassinate Hitler. But now, some cabal of very determined individuals are prepared to experiment on your brother, should they manage to catch him. It's all come down to these days and nights, these hours, David. That's my only plausible reading of all this. It really doesn't matter whether my field and your field coincide. Nor did Einstein imagine Hiroshima. I don't give a damn whether the world is about to end. Or whether your research and family history have so thwarted your sense of adventure... No, I didn't mean 'adventure.' I meant..."

"Forget it. I know what you meant."

"We have to move. It's getting late," Taman said.

They resumed their expedition, albeit with terribly labored effort, through the insanely difficult habitat.

At some point Taman stopped and declared, "Of course it is radical. It sounds... you know how it sounds. David, I've been hindered, curtailed at every juncture, to be sure, in trying to do the right thing. My scientific curiosity is pronounced. But I also want to save him, just like you do."

"Not just like me..." Lev was swirling. "I've got to sit."

He plunked down in the snow. Taman sat beside him.

"Are you all right?"

"I don't know, Taman. I don't feel well."

Taman felt his forehead, then his pulse.

"David, your pulse is way too high. And you're burning up. I need to get you out of here."

"No." And he placed to nitro tabs under his tongue.

***

"Oh, shit. What if he dies out there?" Claire said, turning to Hobbes.

"We're not seeing any of this. We don't know about any of this."

"Oh come on, Allan."

"He's right," Jake said.

"Of course I'm right. What would you propose, anyway?"

Claire could not be more lucid in her thinking at this point. She'd already wrestled with too many demons bearing Belarusian infestation, and suffered far too much in-house contradiction ... spying on our own ... invading privacy at a level that made "1984" look like Disneyland.

"We call Henry at the State Department."

Chapter 79
The Marriage of Figaro Factor

They sat for much longer than Taman felt was advisable, but he had no intention of pushing Lev's heart, he'd admitted, was pumping furiously to keep up.

After a while Lev suggested, "If you're really able to find him, Taman, it has to be my decision what to do. That is not up for discussion."

"I understand."

"Do you? Do you really? Tell me that you truly get it."

"It's dark. Let's sleep here tonight, try to sleep here. Tomorrow at dawn, you make the decision. Is that fair?"

It was their fourth night out in the storm. By his vague tabulations, Lev had been in the country of Belarus for eight days, and in Poland for some 36 hours, not counting his time in transit at the Warsaw train station. But even these hours were blurred. Taman himself felt that he was reaching his limits.

They blew out their one candle and got comfortable, Lev in a very ancient relic of a sleeping bag. It was American, with the name "Holubar" written on it. Taman, who'd gotten it from eBay, claimed it was the best bag ever designed. He was right. Lev actually slept.

The wind continued now as more of a lullaby than an offensive. Taman envied the fact that Lev fell into a most peculiar dream almost immediately. Pronounced alpha state snoring – it certainly kept three people awake at App Opps.

And, Taman knew, it is easy to do so on deep snow, sheltered by enormous trees. This is how – absent the sleeping bag – humanity might have evolved, he had come to believe.

***

June, more or less 2:20 in the afternoon. 1938.

The boy's birthday was soon to pounce upon the world and his family had made much of it in advance. Now, in his favorite woods a few miles from their little house, he lay beneath a gigantic tree playing in the grass, following butterflies with his palm. Just the two of them.

His older brother was there with him. Together they lounged with their backs up against a great old oak and counted the number of bird-calls. A spotted eagle, a white-backed woodpecker. Flycatchers in abundance.

A beetle crawled through the grass beside them, struggling with each blade as if it were a great alp. The beetle seemed determined to explore each summit, and then move on to the next one.

Simon put his face into the grass, giggling. And then did something for the first time in his life: He pressed his ear against the base of the tree.

What are you doing? David had said. *Listening*, Simon replied.

To what?

Don't you hear it?

There were majestic, biblical clouds in the sky, David recalled, frilled and spotted with colors of blue and mauve and shamrock wisps of gold blending with the brilliant sunlight.

That's when the world turned upside down, he thought.

A rumbling onslaught, the sound, huffing, puffing, hooves restless against dirt, a thunder subdued in the breathing of something gloriously... Other.

Hidden just inside the distant forest verge could be heard the shuffling of more than a dozen wisent on the move. And then they all stopped to wonder, directing their vision out into the clearing, gazing with precise vision at the two little people barely populating the meadow within a mottled shade.

There! Simon had gulped, eyes startled, silently pointing to a spot not fifty meters from where they sat up.

A wisent calf, her mother just behind, stood peering at the two bipedal little clumps of life that were clinging to the huge tree, as motionless as the grass upon which they were, in a way, pinioned.

In two weeks time, David would be leaving this world, this heaven of ancient forest, so blissfully indifferent to the forest and the boys, to centuries of insults and scourges, for a place called Ireland. He had no idea what the trip meant: a vacation with his uncle, that was the idea. A vacation. Nobody really knew the word in the Jewish world in which he'd grown up.

He'd looked up Ireland on the map. Far away. Countless unimaginable things. A book of "Kells." *What are kells, Papa?* Nobody in the family knew. But nothing, in sum, to compare with that rare moment when all the wisent stood staring gracefully, gently, at the two brothers.

Simon began to sing. David tried to stop him, certain it would terrify the monsters. But neither his cupped hand nor warnings would stifle his younger brother, who sang melodious tunes from *shul*. He had a lovely voice.

And the wisent stepped forward, one by one: first the calf, then other calves. And then the adults followed.

Until they were so close, David sat frozen with fear.

Simon, on the other hand, was listening to them. He swore they were singing back, but David heard nothing. Only his own heart pounding wildly. That and all the crickets and toads whose added sounds made for something symphonic.

***

© M.C. Tobias

He remembered it all, now. That day.

As the two men on foot slowly crossed the black frozen bog, covered in sleet, tiptoeing as if on a highwire. David had re-assured Taman he could make it and would not crap out on him, like, having a heart attack.

To all sides, poplar and oak, maple and all those shrubs associated with the primeval *Querco-Carpinetum*, leaning in vast arches from what must have constituted billions of tons of snow.

"I've never seen it quite like this," Taman said. "Freezing, melting, freezing, hammered, battered. Amazing! Beautiful."

David was terrified. "I need your help." He couldn't balance on the ice.

Taman went for his left hand. "No, take my right. My left shoulder hurts like hell. That's where the bullet was. Hold me!"

Taman carefully helped the professor maneuver the skating rink. Lev also used his two ski poles.

Suddenly, Lev heard a call he remembered from childhood. "Don't say a word!" He molded the words, as if constructing them from the very frost, frantically.

And then, there it was: "*Keyuke, keyuke, kweee…*"

"What is it?" Taman whispered, kneeling.

"Say nothing."

Lev saw it, flitting behind a trunk of an oak. "That's the one. They're critically endangered." It was right in front of them.

"What species?"

"A white-backed woodpecker, *Dendrocopos leucotos*."

"Oh. That. I see them all the time," Taman said without the slightest astonishment. "They are common."

"Wrong, you idiot. Only here. They're nearly extinct."

"Truly?"

"Yes. Truly. Don't let go." Lev was still teetering. It was all too much for him. "I need to rest."

They found a location and simply collapsed side-by-side into the deep snow.

Lev looked up: "There! My God, isn't she something?" And Lev quietly repeated to himself...*Dendrocopos leucotos*... Lev was breathing heavily.

"How do you know it's a she, by the way?"

"Females sport a black crown. Aside from 80 years of bird watching, research into sexing with wild pairs and color morphs, we also rely on Avibase, our worldwideweb of bird status in terms of comparative anatomy, population dispersion and so on."

"What does it mean, the Latin you said?"

"Tree. Near-Passeriformes, white; family: Picidae; tribe: Dendropicini, I don't know. That's pretty much it. The point is, did you hear how beautiful her voice is?"

"I did. Lovely."

"Mushrooms don't sing like that."

"No, not like that. They sing differently, but one needs to study their acoustics for decades, to listen; that's what I was trying to tell you when you had your shit fit yesterday."

"What? What yesterday?"

"The singing."

"Yes. You are right, Taman. I was wrong. If the birds can sing, and, of course, hundreds of millions of us humans delight in their music, then okay, why not bugs, the grass, the roots...everything you were saying...your mushrooms, the wisent, the red squirrels – they all sing. And yes, my brother has always heard their music, and sung back to them. That would explain his survival more than anything, I believe. Call it the 'Marriage of Figaro factor'. Or whatever."

"I like that. A lot. Orféo."

"Yes. Orpheus. And here we are. Bird-watching in the underworld. By the way, Taman, for whom to my amazement they are common, do you see them this time of year, with that much plumage?"

"Plumage?"

"Jesus Christ. Were you not paying attention?"

"I don't remember. I never thought about it. I'm usually aimed downward, not up. David, quiet!"

It had come again, the roar of rotors high up in the sky, arriving from the Polish side, Taman knew. This time there were two of them. Whispering, ducking for cover, "Those are big helicopters."

Lev watched as all three of the birds flew off in a flash.

"Shit," he mumbled.

***

79 The Marriage of Figaro Factor

A mere eight hundred feet from the two scientists, Ivan Petrovsky had just settled in when he heard the two military helicopters coming. What concerned him was the fact that there had been one, but now two, which he understood to suggest that men might try to rappel to the ground from one of them, notwithstanding the weather: Winds exceeding fifty knots, he figured. They'd be crazy, but then, they *were* crazy.

And probably tired of too many unanswered questions, too much uncertainty.

He had the latest rifle-mounted thermal optic. He could see the blizzard slamming the first fool out of the M135. His 11-millimeter perlon rope was at a sixty-degree angle to the helicopter and everyone seemed to be screaming at everyone.

They hauled him back in and the two choppers – filled with soldiers, he could easily determine – swung away ferociously.

At that moment, with the choppers disappearing back in the direction of the research station, Petrovsky heard something else that was rare in the forest: a vibrating cell phone.

He swung clear around, scoped through the impenetrable ice-caked thicket, and there she was, Ulyana, speaking to someone in an inaudible tone.

***

Norwich had joined Hobbes along with Jake and Claire in the app opps room.

"Henry pulled rank," said Hobbes. "We don't know the real story, but the U.S. Consul-General in Minsk appears to have backed off whatever was going down."

"Guys, listen," Jake said adamantly. "Someone is speaking to her on her iPhone."

"Tell me you can't hear that?" Norwich asked, stunned.

"Yes, the signal is amplified twice, in the chopper, and there, on her iPhone."

They all saw Ulyana, but had lost track of the second person, indeed forgotten him entirely: Petrovsky. No one had any idea he existed.

"*Vy pavinny adstupić. Ja paŭtaraju, plan B. Prybiarycie svaju zbroju i vyjsci. Ciapier.*"

Jake sent the screen shot to an electronic translator and in less than thirty seconds, a robotic voice repeated the transmission: *You need to back off. I repeat, Plan B. Put away your weapon and get out. Now.*

Chapter 80
Daybreak

They'd kept close quarters throughout the stormy darkness. The snow had not stopped. Possibly another two, maybe three feet fell.

"This is historic," Taman declared, his voice hoarse.

"You're sick," Lev said.

He touched Taman's forehead. "Yeah. You're burning up."

"Recurrent bronchitis. I get it all the time. And the accompanying fever. I've never been out here in such conditions, and never for this long. I had TB as a child. Scarring on the lungs."

"So that says it all. You're a wimp."

"Apparently we both are. You should see yourself."

"No thank you."

Both professors had used up the last of their antibiotics and a quantity of other pills and Fishermen's Friends.

"You have any moleskin with you?"

"Yeah."

Taman used the scissors to cut squares, then helped Lev untie frozen shoelaces, get the boots and socks off from under frozen gaiters, apply the moleskin to very ugly black blisters, then get everything back on before his toes began to freeze.

"Your hands are shaking, Taman."

"Yeah. I'm sick."

As they prepared to carry on, Lev had only one question for his ailing companion.

"How did they find us?"

"They haven't."

"But they're closing in?"

"They are flying grids, searching for any movement, using I suspect highly advanced GIS devices, the way I know you and your colleagues might measure for chlorophyll. They are no doubt following our heat signatures."

At App opps, Claire looked down from the pictures before them... *good God, this is so sick*, she thought, giving Jake that look that he knew well. He'd fallen in love with that look.

"What do we do?" asked Lev.

"Nothing we can do. Just try to find him as quickly as possible. You have a better idea?"

"No. I mean, I don't."

They started off in a direction that made no sense whatsoever to Lev. After several hours he asked, while trying to catch up with Taman, who maintained strict vigilance over his increasing propensity to cough up phlegm, "Are we still on the Polish side?"

"We're in the middle," he said.

"How do you know that?"

"I just do. If we had to make a beeline back to my lab, in these conditions, nine, ten hours I would think. We could certainly do it in one day."

"That's comforting," Lev said, although he doubted he could make it.

"It does pose a logistical truth," Taman said. "If, for any reason, you should decide you'd like to try and coax him out, they'll be all over us."

"What did you expect? We're not bringing him out, Taman. That's never been my intention."

"Okay then."

***

When it seemed least possible, frigid atmospheric conditions plummeted even more severely than before, literally encasing the two men. They could not continue.

"We're here, for now," Taman said. His throat was raw but the Cipro was working its wide-spectrum magic. Lev felt Taman's forehead.

"I think you no longer have a fever. Of course, my hands are frostbitten."

"Thanks for that." Taman managed a smile. "Here," and he took Lev's hands and gave them the "dutch-rub" treatment.

It helped.

"How cold is it?"

"I would say sixty below."

"Celsius or Fahrenheit?"

"Celsius."

Lev did some mental chess, then came up with -76 Fahrenheit as the equivalent.

"Really?"

"I think so."

"There's a test... Admiral Byrd who came up with it."

Lev fumbled clumsily to unzip his pants, under his hip-length gaiters, then peed.

"That's strange," he said, quickly zipping up.

"What?"

"I thought the urine was supposed to slow down before it even hit the snow."

"I don't know," Taman said, his voice quavering. He was nearly delirious, on the cusp of the kind of reverie that kills you.

"So what do we do?" stammered Lev.

Taman tried to answer, but could not. He pointed to his throat and just shook his head quickly. It was too painful for him to speak. He did not want to waste any energy trying to do so. All of which underscored the Belarusian scientist's admiration for Professor Lev's extraordinary tenacity.

Этот человек никогда не сдаваться, he thought.

And it was true: Lev came from a family, and from a people, that never did give up.

Chapter 81
11 a.m., Tenth Day

"No more Cipro, my friend. It's gone," Lev said.

"Those Friends of Fishermen?"

"Long gone. Sorry."

They both looked out from their makeshift snowcave. The unprecedented storm had given their desperate bivouac a poignant bathos.

"Do you see that?" Taman's voice rasped.

"Fucking glorious!"

It had quite suddenly stopped snowing. Instead, the entire world was haunted by frozen fog that lay in wisps moving almost like some artificial smoke machine, thickly accented by what could only be described as the most acute, frozen silence either men had ever experienced. A silence that transcended the absence of sound.

Not a bird stirred. The cold was so intense as to hold fixed every gorgeous giant of ice upon the trees, all ready to topple from their tentative reality.

This landscape of vertical ice was something new on earth, Taman felt; an aching precursor, vivid windows on the future, for very few.

For hours they lived amid the frightening stillness, fearful of everything, waiting… wondering when the next chopper roar would break the spell. But no chopper came. Taman understood: It was just too cold for any manned vehicle to brave it.

But something did happen, and neither scientist had the expertise to delineate one drone from another, but it was a drone, and it came through the forest, about fifty meters above them by absolute surprise.

"Duck!" Lev managed to grunt.

It passed by them, not to return.

"Did it see us?" Taman asked the Professor.

"Undoubtedly."

"Then that's it. We have to try."

By that, both men understood. If Lev was to ever verify the life of his brother, to see him alive, if, indeed, he still was, this was the day.

They got their gear together and proceeded. It was foolhardy, and they accomplished one step at a time, until Lev realized he was also out of nitroglycerin tablets.

"I'm a dead man," he uttered, totally out of breath, leaning against an icicle for support, on the edge of a frozen stream.

"You're going to make it, Prof."

"It's already getting dark."

And that's when it happened.

Chapter 82
The "Parthenon" of Bialowieza

They had reached an upland bank, beneath a vast canopy of a single tree, its base joined in force with several others whose combined and intermingled taproots spread far beyond and beneath the creek. Taman knew the spot. He had been there two years or so before, in the late summer, when it swarmed with encephalitis-inducing ticks, and buzzed, Scriabin-like, with horseflies and mosquitoes.

The Belarusian turned ever so gently – relieved, actually – and motioned with his gloved finger to his frost-caked blue lips: Shhh…

This was no woodpecker.

David reached his left shoulder side, out of breath. He looked up through dense frozen fog and saw what Taman was referring to.

Above them, ghosts of enormous bulk and palpable breath moved secretly, one step, then halting, then a step backwards, then a turn, followed by a freaky mass motionless gaze. Sovereign stares, plural, adjoining imperial fixity: marmoreal and grave.

Everyone seemed to be standing on glass, the world ready to shatter.

Taman grabbed his binoculars and counted.

Whispering to the max, "I've never seen so many, not even close, not at once. This is unprecedented."

David could not get out his glasses from the top of his pack. It would have most certainly made too much noise. "How many do you see?"

"Sixty-two, and something else. *Someone* else. David…"

"Sixty-two…." Lev said. He was breathing heavily, exhausted, holding his waist with both hands, his head piled forward, his mind minced, bitterly cold, everything caving in. His lungs were bursting and felt like dry ice. He knew a heart attack was on the cusp of his last hours on this planet, in this mode of bare being. As if no oxygen were binding to his blood, or cyanide had entered his final moments.

But his mind seemed utterly preoccupied with something other than self-preservation.

"Sixty-two?"

"Yes. He's there, David."

"Sixty-two, you are certain?"

Taman systematically counted. Then reiterated, "Yes. Precisely sixty-two individuals."

"Quite remarkable."

"What?"

"The Parthenon."

"What are you talking about?"

"Taman, the very one atop the Acropolis. Eleven columns on the two long sides, eight and an inner six on each of the short sides. Sixty-two Doric columns. We all learned that in math class. Didn't they teach you kids that? The Golden Rectangle. The Golden Ratio. And paradise, so to speak in the middle, paradise...." And Lev's thoughts racing like synapses jamming in some electrical transmission meltdown.

"...Although history tells us the Parthenon actually referenced, in the ancient Greek language, to the apartments for unmarried women."

Taman realized that his companion was hallucinating, hypothermic, trembling.

"David," he said, "listen to me: There's a person at the site of the Golden Rectangle. Not a woman. Don't move."

The air was crystalline and steely, as fragile and ambivalent as mica, transparent and razor-thin in its ferocious chill. An hysterical level of cold that had the freezer perfection of a bullet traveling without definition, at a speed beyond all ken, like an airline crashing into the sea in bad weather, everyone aboard screaming and silent, at the same time.

Even the unimaginable hiss of superheated plasma as a Leonid meteorite shower illuminates the night at 4 kilometers per second would not have sounded like this.

A stillness of absolute zeros counting backwards into the eternity of cold separating the two men down below the huge herd of wisent that had appeared on a rim, forty feet above the icy bog, grunting, as it were, in a language of soft muting vocables; all in a frequency beyond any possible translation towards human conditioning, training or understanding.

Transcendence – pure, simple transcendence, by any other name. In fog, the floating hues and textures of a Franz Klein blue.

Save for one individual, visible through the frozen atmospheres in Taman's binocular perspective.

Breath both frozen but slowly hovering... Standing in the center of that bovine bubalus connection dating to the Himalayan yaks at least nine million years ago.

There amid the *Bison bonasus*, who – snorting, as it might be described, or humming in a frequency we are unaccustomed to, but perhaps Monteverdi, or David's Mozart knew perfectly well, with the normal course of communications over the last several millions of years, punctuating the gap of oxygen, a trail of breath signaling the design that might yet be defined as a boundary layer, like those reckoned upon in cosmology – was their orphan singularity at the end of time.

A human.

A black hole in which all goodness, regression, progress, hope, destruction, mindless sub-atomic particle traffic moved, got motionless, shattered beyond all descriptive faculties, created and destroyed universes. Made human history in an angstrom of a millisecond of dust and called him Hamlet. Or Socrates. Or someone like Wittgenstein, or Robinson Crusoe.

Denominated it by a lamb as rendered in the Van Eyck Brothers's "Altarpiece at Ghent." Or in two fingers atop the Sistine Chapel. A white dove. A manger. A Torah. Linear B. A handprint at Altamira or the recently discovered petroglyphs near Burgos in Tamaulipas State in northeastern Mexico. But a lamb who had been hunted, unlike those described by the German composer J.S. Bach during a concert the great composer had rendered for an Easter, centuries before. This lamb was on the run from twentieth-century history. A terrorized lamb.

Ungulates at the mammalian apogee of terrestrial evolution, four-legged, but for one demonstrable discontiguity.

Then David, glasses affixed, saw him as well, squinting up at the astrophysical haze. Four legs, four legs, four legs, a Parthenon of sixty-two separate four-legged bovines in formation, atop their wooded golden stylobate.

"Yes. It is sixty-two. I count sixty-two."

"I told you, sixty-two. Amazing."

"That's not what's amazing."

For in the heart of their private, pensive, sub-family protectorate, there, unmistakably, although hiding fiercely, was someone bearded who was...yes, most definitely: two-legged.

Chapter 83
A Single Gnarled Gutteral

Taman looked down, inched up his parka over his sweater, woolen shirt, T-shirt to make absolutely certain his geocoder – which would, otherwise, have provided a precise GPS position – was turned off.

They were approximately nineteen kilometers, he knew, from his research station. Five minutes flying time by military helicopter. Days, if men and Japanese Tosa Inu relentless and bristling tracking canines without knowledge of the forest tried to find them. But the precariousness of the moment was beyond any known barrier-effect.

David started up the slope, slid back down and choked on a curse, coughing uncontrollably – he and Taman, both, a fine team at that moment – "We're pathetic," Lev mumbled – holding back the terror of breaking through ice into the deep thick of brown water. "Help!" he tried to shout, the frozen water asphyxiating his tongue.

Taman got to him – he had done this once before, with a brother, ice-fishing in Central Belarus just two months before his military duty – hauling out and steadying the 87-year old, bringing him to his feet, off the ice and back on to the bottom of the black-ice slope leading to the ridge where the bison now displayed fast-growing anxiety.

The two-legged member stepped forward from the herd and leaned out from the rank and file, his look frozen mournfully at the two humans down below.

An unmistakable caesura, a grasping pause in the lifecycle of two siblings.

David tried to make some meaningful contact with the noble and suffering personage who returned the penetrating gaze. Both were trapped in something as distant and aloft as a star that has imploded, and has some unusable name in cosmological circles. It is not the stuff of the great aesthetic jubilations sent back in code from an orbiter amidst the moons of Jupiter; no euphoric reunion of the long-lost; no enigma-code break-through.

Rather, a gnarling gutteral that shattered the burning cold of near dusk: "GYYN `AKK`q! LAZ MIR ALEYN!"

Several of the wisent retreated, snow puffs rising above their disoriented hooves.

The voice was no declarative roar, but the weak, near extraterrestrial appeal, as it were, real enough. The horrifying and ambiguous utopia of an exhausted and desperate man's cry.

"That's him. That's Simon. What do we do?" Taman whispered frantically.

Lev's heart appealed, to the air.

"David, what do we do? What did he say?"

"To leave. To get out. Now. To leave him alone."

The herd, and the two-legged member of that herd, were backing away into the impenetrable whiteness of white forest.

"There's no time, David. What do you want to do?"

***

"What now?" Jake, tensely methodic, blurted to Hobbes, who still sat in the room with him and Claire. The three of them had worked in shifts, allowing four-hour breaks for sleep, and the odd, rapidly gobbled burger.

Jake's parents, who lived all of ten minutes from them in Fairfax, were looking after the girls.

"We do nothing," said Hobbes. "It's always been GIS reconnaissance, security, if you prefer. It is not an international incident. It is not our jurisdiction. We have no authorization to do squat. If an American citizen needs help, he goes to our nearest consulate. Nothing more to do."

"Jake!" Claire cried out.

"Zoom out," Hobbes declared. "Who the hell is that?"

***

The Belarusian military choppers had both set down at a distance equivalent to three minutes of hauling ass in this weather from the research station, in a clearing sometimes used as a feeding station for wisent, favored by foxes and wolves. Voles and mice loved the seeds of the meadow, especially after humans left hay for the wisent.

Now it was the site of an all-out invasion.

John Vespers, ducking under the illusion of risk from the rotors, shouted to the General, "Are you prepared to send in your troops?"

"No. We wait."

***

At that same instant, two individuals, a mere 300 feet apart, separated by treacherous hurdles, but – at that moment, a clear line of sight through the forest – saw one another.

Ulyana dove for cover, as an arrow came flying at nearly 1.4 kilometers per second, missing her head by the eleven inches separating her cranium from the near base of a giant Norway spruce, into whose wintry bosom the 300-grain carbon arrow shattered the icicles with explosive kinetic force.

She whipped around but it was too late for her to see who was pursuing her.

The stampede happened instantly.

The two scientists stared at the vanishing. Within seconds, they were gone. The herd had fled, and with them, Lev's brother.

Had he and Simon made eye contact? He wasn't sure. Had his brother recognized him? No way to know. Was there anything else to do?

Scenarios converged like a sub-atomic particle panic at the Hadron collider, smashing and breaking, exploding and recanting.

David Lev was lost. Destroyed. Reborn.

A total modification of reality that just attacked every nerve cell in his being.

But none of this constituted pain. The camps, in summer, in winter. That was misery. Absolute pain.

And no one could speak of the camps, not unless they had been there at the time. No one, but those upon whom human vagaries had inflicted their worst cruelties, would ever have the right to represent what happened.

For the rest of us, thought Lev, all else demanded adamant memory, silence and prayer.

In days to come, he would remember that stare. How Simon had sadly gazed back, down, at him, across the eternity of a tannin-filled frozen bog in whose nameless molecules and ecological facts of life, two separate worlds lay in motionless, asymptotic contact, memory, graven in frost.

That unclear ferocious moment would reverberate over and over and over again for one David X. Lev.

Chapter 84
Escape from Nature

But for now, desperation was the only *modus operandi*. And David Lev, for all of his years in the wild, his exasperating emphasis on precision, and Latin, Greek, ornithology, quantum mechanics, the end of the world, the beginning of the world, the author of the definitive text on ecology, eco-dynamics, over-consumption, world-population, the coming climate catastrophe, extinction upon extinction... the Rio Codex (as he had described the insoluble chaos to his wife, Sasha) etc., etc. – for all that, he was cold and tired, and felt like an absolute novice at life.

All this had nearly struck down his Belarusian colleague, who saw the unfoldings and rapidity of life going down, going away, forever, as if in slow motion, the entire beginning and end of human history in the personage of his companion, David Lev. But also his own family saga, the memories of his father, the knowledge of all those in his family tree who had been martyred to all the horrors of human nature gone awry, gone forever. Leaving nothing but the further horrors of memory.

He still had, somewhere at home, Sarah's copy of *The Road to Utopia*.

Both men were helpless, in that sense, to do a goddamned thing about it.

None of the insights of such great war historians as William Shirer or Brian Reid; Sir Fitzroy Maclean, 1st Baronet, or Šemso Tucaković could be appealed to, flashed Lev, to offer up any blueprints for making sense of this concavity of family perturbations in the thickest turmoil of the century.

There was Lev's brother, fronting full independence. Beside him, an entire tribe of critically endangered wisents. A multi-generational intimacy, trans-species acceptance, preeminent mammalian nurturance he could not have believed, no matter how many Ph.D.s in sociobiology, taxonomy, evolution, anthrozoology, anthropology... Lev was lost, like the day he was born.

He could not save his only brother, and his brother could not save David Lev.

Indeed, that was not the question.

Like two distant universes, unspeakable parallel tragedies spanning two centuries, there was no communication, other than those five unmistakable words shouted out in angst, true torment – not existential sorrow, not some erratic emotional outpouring of woes or long-suffering plaint.

Rather... the harrowing, mystical perfect truth.

"No. We leave him. There is no responsible, humane alternative. We must get the hell out of here, now."

"David, one chance in your life, this is it. Are you sure?"

"Listen to me," Lev declared resignedly. "We have to leave, right now!"

***

They made a rough bivouac, then left by dawn. It took them most of the day. They encountered no one, but neither of them for one moment truly believed it could end quietly.

By 4:30 that afternoon, they'd gotten back to the station. As they expected, they were not alone. At least twenty military had made themselves at home. They were everywhere, but Taman could see that they hadn't touched anything. Odd, he thought.

Taman had prudently stashed his illegal firearm and geocoder in a safe hideaway before breaking camp that dawn.

But they were the least of his concerns, although the gun alone – inside a national park, despite who he was – could have cost him two years in prison.

"You must be Dr. Lev?" the top-ranking officer said. "May I shake your hand, Doctor?"

"Nice to meet you too, General."

Taman awkwardly shook the broad shouldered wrestler-type's hand, when it was extended, as well. A very strange gesture of reconciliation, but then, the forest was full of surprises.

"How was the bird-watching?" the General asked, with an air of genuine interest.

"Saw some lovely woodpeckers," Lev said, out of breath, guzzling the last of his supply of water from a plastic bottle.

"There are many woodpeckers all over Belarus. Very common birds. I used to delight in seeing them as a child."

"Well, what do you know."

"Why don't I make you both a nice cup of tea with rum? Sit down, relax. You must be exhausted. My God, your Consulate has been so worried about you, Doctor. At your age. In the storm of the century. No wonder you were both lost. Did you realize that you crossed into Poland? Ah, but never mind. Such dreadful visibility. High winds. The temperature inversion. I have never witnessed such extremes. Even our search and rescue helicopters were useless. The main thing is, you're alive."

"So far."

The General said something rapidly, in a crisp military Belarusian, to Taman.

Two days later, Lev was on a flight out of Minsk to London, then on to Los Angeles.

Chapter 85
Paris

Skype. 11 p.m., Paris time. "Hello, David."

"Sorry, no video. I look like hell. You don't need to see me. Moth-eaten PJs. Don Quixote, as it were, à la Doré. No matter. Where are you?"

"Sofia and I are staying with a relative in Paris. On the Rue des Rosiers."

"Rue des Rosiers, the 4th Arrondissement?"

"Yes. Quite close to the synagogue at 10 rue Pavée."

"Very nice. Why Paris?"

"Had to go somewhere."

"They let you go?"

"Yes."

"Good choice."

"I've been once, long ago, to Nice. I like France. Anyway, I was too problematic for the system."

"Is this cousin related to the one in Rio?"

"No doubt. We're all related, you and me included. I think that's been ordained."

"St. Ignatius of Spain – the one intent upon improving the message of Saint Francis, had a first cousin in Poland, a saint, of course, who was no doubt related to a descendant of someone with the Society of Saint Mary, who kept secret her nephew, one of those well-connected spiritually hybridized types who studied in the local *yeshiva*. Sound about right? One big happy Renaissance family. Jesus was a migrant worker, did you know that? I read it on some Guatemalan protestor's placard seeking sanctuary in Oakland, near San Francisco. We've all been fucked over. That makes us more than cousins."

"David, you sound like you're on drugs?"

"I am. All the pain killers I can get."

"Understood."

"And this Parisian relative, also a bartender?"

"No. She's a librarian."

"So I can expect to read about a terrorist attack on a library in Paris, something like that?"

"Don't even say those words."

They were silent for some time. Sasha, who had been listening half-attentively, returned to sleep.

Taman asked, "So how's your heart?"

"A few months older."

"Good. Listen, we need to speak in person."

"I have nothing against Paris, believe me, but why now?"

"There were numerous and awkward questions by the authorities. Weeks of polite interrogation not meant to be construed as such, of course. But that's what it was and I went along with it to stifle the possible afterlife of the forensics."

Much more silence.

"Got it."

"So when can you come?"

"Give me a week or so. I need to do some things."

"Well, on a cheeky note, I'm unemployed. We're here on a three-month visa. After that, I don't know."

"Don't worry about it."

"Easy for you to say."

"Seriously, I can easily swing something for you in the U.C. system. Fill in where I've left off, so to speak. Interested?"

"My mother, as you know, remains in Minsk. My two kids. Honestly. I'm tired. The thought of resettling."

David could read between the sighs.

"You're way too young to be tired. Anyway, not resettling. An Adjunct Professorship in biology, something like that hold any appeal? One week a month? Build up lots of frequent flyer miles?"

"Go on."

"No preparation needed. They know nothing about Belarusian ecology in Hollywood."

"Hollywood?"

"I don't mean, strictly speaking Hollywood, Taman. More like West L.A. Just Google Earth it."

"But close to Hollywood. Sarah loved Hollywood."

"Well, yeah, twenty minutes by car."

"Maybe, of course. Hell, yes! And there is forest there? Some mushrooms maybe?"

"You're being humorous. That's not like you."

"No parks? Come on."

"Yes. Parks. A few. Just like in Minsk. But warmer. The warmest ever. I'm sure you'd love it. And there's a mountain range in Los Angeles. We've got mountain lions."

"Seriously?"

"Just look it up, for Christ's sake."

"My family is struggling, David. Belarus is struggling. Oil prices are nearly half. Tied to Putin's policies. The ruble keeps falling. There is no future, although we're better off than most, I suppose – I mean, personally speaking. And I must say, Paris is everything it's cracked up to be."

"First time in Paris, huh?"

"I know. Ridiculous."

"For a half-Jew, the Marais is definitely the place. You can get half-kosher croissants, as I recall. And the newly done Picasso Museum should be open soon."

"We'll go. I've been to the Museum of Natural History. The books are mostly in French."

"Well what did you expect? Buffon wrote in French. Tell me you don't read French?"

"Why would I read French?"

"You know ten Eastern European languages. Get busy."

"And what about you? After so much?"

"The timing was perfect. Ventricular tachycardia. But we were literally pulling into our driveway after getting home from the airport. So, another eight minutes to the E.R. All dressed and ready to go, my wife driving like a lunatic. I'm gonna be 88. What do I care? Still walking every morning. Knees, shoulder, blood pressure, all fucked up. But we're alive, you and me. Sasha, too, by the way. I want you and Sofia to meet her."

"So, you'll let us know where and when?"

"You and I got pretty good at that, I'd say."

"Yes we did, Professor."

"How is Sofia taking exile? Can they actually do that?"

"Yes. And it was hard when it first got handed down. They questioned her, thoroughly. Quite humiliating. But we got through it. Now she's positively made for Paris. Of course, we're immigrants, temps, no money. Enough for a flat week-to-week, French fries and salad. Pretty exciting."

"You can't love Paris too much. We'll meet at a café. You still on gmail?"

"Different account, here."

"All right. I'm writing it down. Just a minute, Taman. Don't go anywhere." David scribbled Taman's e-mail account on a Post-it, same size as the one that first triggered their unusual connections; he swiveled away from his desktop computer next to the bed and put his arm around his wife.

"Sweetheart?"

"Hmm…" she murmured whimsically, sleepily, dozing off in sync with mnemonic utterance. She'd already drifted towards her own world, as was her remarkable penchant.

"Feel like a trip to Paris?"

And then, in a radically different tone and turn of moods, removing her French wax earplugs in a kind of panic, "What's wrong?"

"I don't mean right now," David calmed her, gently stroking her long hair.

"What's happening in Paris?" she perked up.

"Everything."

Chapter 86
Les Deux Magots

"It's at 6 Place St-Germain-des-Prés, Paris 75006," Lev informed their taxi driver.

"My friend, for 120 years, the center of Europe. You think I don't know where Les Deux Magots is?"

"I'm sorry. Where are you from?" They were taking the taxi from their hotel in the 4th Arrondissement.

"Ethiopia."

"I've seen your wolves, or endemic fox, you might say, in Northern Ethiopia."

"Really? No way!" the driver was suddenly a child overwhelmed with excitement.

"Yeah. Near Gondar."

"Precisely right. Our Simien National Park. World Heritage, you know."

"Beautiful place."

"And you saw our baboons, and colobus monkey?"

"Met every one of them. Know them by name. Even went to a marriage."

The driver was laughing. He got the humor, then elected, "Wow, you know more about Ethiopia than I do, Sir."

"Unlikely."

Sasha smacked David's thigh. "Cut it out."

Lev paid the thirty euros and they headed inside.

"Thirty euros. I remember the days when a taxi in Paris would have charged, at worst, 5 francs for this ride," Lev grumbled.

Inside, there was a swarm of lookie loos as well as authentic pilgrims out for a good time, and a little old-fashioned Parisian nostalgia in the flesh.

"There they are," David said, sighting Taman and Sofia seated at one of the famed mahogany dinner tables with its moleskin chairs.

"Taman, Sofia!" David hugged them both, then introduced his lovely Sasha.

They were at once family.

"I've heard so much about you," Sofia said bashfully to Sasha.

"Oh dear," she groaned.

Sofia smiled knowingly and they all sat down.

"You know, I used to frequent this place," David began.

"When was that?" Sasha asked, a bit skeptical.

Taman turned his focus towards Sasha and asked, "Mrs. Lev,"

"You must call me Sasha."

"Sasha, is there any place your husband has not researched, I mean on the whole planet?"

"Cut it out, guys," the professor insisted.

Sasha shook her head. "No. Although he is presently avoiding the Gulf of Aden."

David was in fine spirits, notwithstanding having been sober for over four hours.

"Over there – see that man! Picasso. And over there, most famously, Jean-Paul Sartre."

"Yes. I understand," Sofia chimed in, on to Lev's indomitable spirit. "I, too, read the Lonely Planet guidebook."

Taman added, "Did you know that Sartre's refusing the Nobel Prize was a big win for the rest of us? The Soviets loved it."

"Yeah, you're right," exclaimed David. "Except I read that he lamented the fact later on. He realized he could have donated some of those winnings to refugees. Just like those displaced persons Orwell had tried to help."

"Very true," Sofia said. She had read all of Sartre, in French.

Taman remembered their trainride out of Minsk. The mystery of that publisher of *Animal Farm* in Ukrainian. The strange coded "jv" never clarified, first e-mailed and received as spam, when both he and the professor were in Rio, and so forth.

The waiter brought them menus. Drinks were ordered all around.

And the more David Lev drank, the more his nerves became frayed, accelerated towards a sweat. He was alarmed that nothing, as yet, had been exchanged pertaining to anything of importance. He was worried about that. Sasha saw it coming on. She knew the signs.

At first, Sofia heard R. Nathaniel Dett over the music system in the café, from the composer's "Eight Bible Vignettes/Martha Complained," a most welcome antidote to the more universally recognized 1888 "Gymnopédies" of Erik Satie, a kind of congenial Everyman tune played everywhere in France, not just in elevators. A mood to lift the human spirit.

Of course, true to her Belarusian roots, Sofia also knew about Satie's sad side; one piano on top of another at his residence within an apartment building that, according to the story, he lived in without ever receiving a single guest for 27 years, dying slowly from cirrhosis of the liver, his domicile in utter chaos, what some labeled "squalor."

Later on, after a few more drinks, Sofia pointed out the opening passages, sublime and forever perfect, of Bizet's "L'Arlésienne," from the first suite.

"Do you know it?" she asked those at the table. "Taman's mother taught me to play it."

That ever so delicate barely present symphony, full of joy and trust, the Adagietto taking its time towards the full light of day. And it was warm.

...Sweetheart, what? I'm sorry, what? This is all a dream... Lev was imagining. Swirling in his head, which he presently steadied. Every nuance of the day seemed to touch his heart in huge accents.

Lev helped himself to a stiff drink, a second, a third – Sasha noticed – maybe a fourth in a row. There was no reason to stop him, she believed. Not at this age.

Somewhere in Lev's brain was fixed that wondrous Henri Bergson's essay "On the Immediate Data of Consciousness," but he had no clue why the hell that had suddenly opened to his far-reaching nerve endings amid the joviality of his surroundings.

Probably, because it was summer in Paris and David Lev was buoyantly beyond critique. Indeed, the case had come to a head: He had some level of dementia not entirely in line with the normal progression. He could still function. Put on a show. But Sasha understood the grief. For that's what it all boiled down to.

They would not, could not discuss it. They had already done more than their share to expose the flaws in human nature; to question the grand staircases of human evolution; to give up on saving the world.

Chapter 87
The Synagogue de la rue Pavée, 4th District

Taman and Sofia were there before David and Sasha. He had time to explore and to consider how he wished to break all the news, but this was definitely the place to do it. And, by remote coincidence, the right day. It was the Festival of Weeks, *Shavuot*, commemorating the giving of the Torah at Mount Sinai. Leviticus, Chapter 23. The 29th verse recognized "the day of atonement."

He and Sofia spent ten minutes or so wandering throughout Hector Guimard's 1913 Modernist creation, mostly utilized by the Orthodox Jews of Russia and Poland living in Paris. It was named "Agoudas Hakehilos," with an open book of the Ten Commandments, as if to anticipate this meeting with Lev and his wife, welcoming one and all into the *house*, with its adorned motifs of various garden vines and plants, a library, beautifully illuminated.

The history books recorded that on Yom Kippur Eve in 1941, this synagogue, and six others across Paris, were blown up by the Nazis; this one was rebuilt and registered as an historic monument in the late 1980s.

By the time David and Sasha arrived, rain was coming down as one might expect in the neo-tropics, but rudely splashing on pavement, of course, giving every passing vehicle an aqueous throw-weight with the velocity of paint-ball ballistics.

They closed their umbrellas once indoors, greeted their friends, then followed them to rear seats within the sanctuary, joining in an existing service, opening prayer books that lay on wooden pews before them. Afterwards, David and Taman had a moment near the Holy Ark, wherein the Torah resides, while Sasha and Sofia wandered upstairs towards the library beneath the skylights.

"You were going to tell me," David said. No resistance to the facts.

Taman took in a deep breath. "Right. By the time we'd reached the compound, another storm was on its way from the Arctic. Ulyana's presence the first two days, nights, made it quite impossible for us to speak openly, as you know."

"But you were clear that it was over for you. You had come, you had seen the situation, and – while I didn't want to probe too much – it was clear to me that your decision was to leave Belarus. Leave the whole amazing, sad, maybe not entirely sad situation...I don't know. I simply saw that you had come to some kind of closure. Am I right?"

"You know I chose to leave."

"Yes. And as for me, it was over. I learned many things with you."

"Stupid, complicated things, perhaps."

"Hardly."

"Taman, maybe peace is now at my fingertips. Insane but true. Frankly, I was surprised by the events upon our arrival back at your compound. I had expected... more trouble."

"You are a famous American. They were not prepared to, you know."

"But that story about rescuing us? What did the General say to you?"

"Many things. But, you see, they wanted it to seem like some heroic search and rescue. Of course they did. They knew, they must have known, that there could be outside surveillance. They have the time on their hands."

"So?"

"Okay. Sometime between our departure from my research lab, your getting on that flight out of Minsk, and now, I was able to return nearly straight away to the forest again."

"You weren't worried?"

"Of course I was."

"And your lab was intact?"

"I'm getting to that. When I returned to the compound – no incidents on the train this time – there was another storm, an even more vindictive one, if you can imagine such a thing. It, too, arrived with a mind of its own. As if a great painter had grown angry with his canvas. It came in like a tsunami. I've never seen such back-to-back chaos this late in spring. I know they are discounting these so-called anomalies in places like Washington D.C., and Alaska. But not here. Not between you and me. We know exactly what is happening to the world."

"Go on."

"I got to the bog. Nothing. But two more kilometers, eventually I found the herd. Your Parthenon."

"Got it."

"They all ignored me."

"All of them?"

"Yes."

David's eyes closed; he took a seat on an aisle. There were other people arriving for prayer and Taman sat down beside him, confining his remarks to a whisper.

"They were protecting him. I think he had an infection. It was low-grade. He did not show himself. I only saw his legs. Good God, he is strong. Like a biblical patriarch."

"Continue."

"They were browsing. I truly don't envy them."

"Right."

"Eventually, they moved on. No stampede. They simply left the spot."

"Go on?"

"Something seemed odd about it. Then I realized. I knew exactly what was going on. One of their own had died, a calf. It got me to thinking."

"About?"

"A ploy."

"I don't understand."

Chapter 88
The Survivors

"I was on my way back to the station to get supplies. A shovel, pick-axes, and so on. Moving quickly, feeling much stronger, when two really ugly voices spat out that dreaded word: 'Papers'."

"Oh, terrific."

"Exactly. One notch before either being shot or jailed."

"But you work there. They know you. We'd already been through that whole pantomime when you and I got back to the station. Saving the poor American tourist and so on."

"These were not park officials, half of whom had, over the years, asked me for a good word, or a job. BNAS pays me better then any of those guys make. No, these were strangers, army, same type you encountered, with none of the politeness. You were gone, remember. This was the behavior you read about from the Cold War period. I saw some of it throughout boot camp during my two years in Belarus military service. The most macho among them get very surly in the cold. I tried everything – the scientist bit, my old days in the military bit. I had everything to be worried about."

"So?"

"They were under real pressure. I could see that instantly. Somebody was calling the shots from very high up. And, I now believe, that *high up* was a state-owned pharmaceutical oligarchy or partnership, as I'd always feared. Remember? My worst premonition."

"We discussed it."

"Yes. They had been trailing me and it was just dumb luck, on my part, that they caught up with me after… the bog. I felt sorry for their dogs, who were flailing pathetically in that deep snow. Especially the German shepherds."

"How ironic."

"Yes. It was a wet, sloppy snow, but many times more impressive than I'd ever witnessed. We made it back to the station and I found myself in the unpleasant company of a dozen military, real dicks. I had hoped, of course, to spend one more night in my own bedroom of many, many years, then, the next morning, go back out and do what I needed to do."

"There was an American there."

"Interesting. Anybody I might know?"

"He knew your name. All the facts of your *rescue*. He asked me many things. Like what I was doing in Rio."

"How interesting is that!"

"What did he look like, just out of curiosity?"

"Tall, blond, perhaps mid-forties."

"Well, what do you know!"

Taman took out a notebook and wrote something down, then handed it to Lev to read.

He asked if I knew where Ulyana was. I said no. I lied. I had found her dead, not far from the bog. An arrow through her chest, two rifles by her side.

Lev paused, handing the little notebook back to Taman. "That restaurant never appealed to me. There are far better ones in Paris," Lev said.

"So, you understand Paris."

"I'm starting to get the hang of it. The good, and the bad."

"So what next?"

"What I had expected. They went through everything, seized the hard drives, confiscated test tubes, the little refrigerators, all of my 18S-ribosomal RNA gene conservations, dozens of clones, rDNA primers, our entire phylogenetic database, who the fuck cares at this point. Twenty years out the window. I remember most of it. But really, it doesn't matter, David. There was nothing important, no science of any consequence, really."

"Well, you always said so. A waste of many years of research. But we both know, good science is tough."

Their bullshitting was well calculated. Neither trusted the world anymore.

"Fuck it, to use Sarah's poetic description for these things. My research could not have resulted in any breakthrough for the world. Let us be honest. Did Galen, or Charles Darwin really change human history? They just illuminated good medicine and the nature of island finches. Certainly nothing, on the surface, of any significant drama. No relevancy except for those few who are idly interested and with enough free time on their hands, to think about it all. Certainly no guarantee of a better, or longer life."

"A damned luxury of meditation, when one comes down to it," Lev added.

"Absolutely. No different from Zen. World War II happened despite Darwin, and even your esteemed Wallace could not change the fact that Hitler showed up. Neither Aristotle nor Mozart could stop him. It happened and it is happening again, and people like you, Professor Lev, have warned us for decades, and nobody but the very few have the time, wherewithal, or emotional nerve to listen."

"So that sums it up, then?"

"Basically, I suppose, there were two camps. The first, those wanting anonymity, with everything to lose. The others, solid capitalists, and why the hell not. Of course, the two goals, in this case, were irreconcilable."

"Look on the bright side. There is always a bright side. The food here in Paris, for example."

"I worry about my mom, our two sons."

"They'll get out. They'll visit you, but Dijon's the place. Take it from me."

"Right, you no doubt discovered some new famous bird species in Dijon."

"It's Paris, without all the hassle."

Both men thereupon sat listening to the cantor, who sang beautifully to a half-full Synagogue. Not half-empty. Lev turned and looked at Sasha, who at that instant recognized a gaze she had not seen in his eyes since nearly the very beginning of their love affair, sixty-odd years before, in Ireland.

But this was no lilting Irish tune from County Derry. No nostalgic remembrance done in the strings of some Percy Grainger, whose music the teenage Lev had known so well. Not even the soothing, if sadly premonitory music of a Yom Kippur service when the cantor of an earlier era had sung his sadly state of glory, circa 1936, in northeastern Poland, the last such service David ever remembered.

David nodded, conveying a clear signal of concord. Both men gently smiled.

As he continued to follow the Hebrew and the music, he focused upon the nicely adorned copy of the Old Testament in front of him, the written Torah – those Five Books of Moses, allegedly composed by Moses himself in 1,273 BCE, a total of 613 commandments, or "*mitzvahs*."

Chapter 89
The Crossing Over

The two couples had had dinner and were leisurely taking in the calm, summery night air of Paris, walking across the Passerelle des Arts, over the River Seine.

This time, it being an early Saturday night, lots of foot traffic. They'd had much to drink. But David had one last critical piece of information to glean from his friend.

They all stopped more or less in the middle of the bridge.

"Taman, I think it needs to be said: Not giving a shit about privacy is one of the great forms of revenge."

"Got it. So I will speak freely, then." He glanced at his wife and David caught her sign of compliance. She obviously knew what he and Sasha were about to hear, leaning against the perfect night, the Louvre to their right, the Left Bank, over there…

"I retrieved a shovel, also a pickaxe, returned to the site. Many weeks later. I did not touch Ulyana's body. I'll leave it to future forensics to figure out. I never did trust her. But, I don't know. It's very sad. Somebody had it in for her. I don't know."

"I'm sorry."

"She was always a mystery."

"Taman, what about the Parthenon?"

"So. He has, by now, lived through nearly four generations of the wisent. That is my assessment."

"Four generations?"

"Seventy-five plus years. I made sure to cover my tracks. To render her final resting place," Taman declared with emphasis, "near to her home – a magnificent dead wood, an oak that probably had a circumference of 240, 250 inches."

"Twenty feet? What are you saying? Her?"

"Yes. Under the freeze, surrounding that tree, vast assemblages of the quite tasty polypore mushroom, *Grifola frondosa*, as well as beefsteak fungus, a quite delicious morel that is very rare, protected, but she was feasting, probably her whole little life. I buried her there. Very deep. The corpse is already half-rotted."

David's eyes jerked, as he tried to assimilate Taman's emphatic ambiguities.

"Taman, I told you. You have to live as if nobody is listening."

Sofia nodded to her husband.

"David, you are absolutely right. Anyway, I can never go home to that forest. So, this is my new beginning."

"Yes it is, kid. Now tell me what happened."

Chapter 90
Coda

Taman handed David the notebook. He had already written it down. Paranoia had long been a habit of Taman Chernichevsky's:

The bones of a wisent calf. She died at the age, more or less, of three months. And I can just see their faces digging her up in summer. A rotting calf. Swatting at the flies. The ticks getting up their asses. Your brother is long gone. He's with his family. I think he's okay. They'll never find him.

David handed the notebook to Sasha, who read the words, as well.

The two couples remained quietly standing in a breeze that had turned up.

"May I ask how old he would have been?" Taman inquired.

"Do the math. Eighty-three I suppose." And after a few moments of not utter solemnity, but a kind of calm, David wondered aloud, "Should I not have come? Or, perhaps, actually rescued him, taken him home? What's your opinion? Be honest with me. Please. I know we labored it to death. But I'm still, well… haunted would not be an exaggeration."

"We could get philosophical. But why do it? I think about what he must have suffered, especially those first days and nights, the early years. But, frankly speaking, having spent so much of my life out there – granted, differently – but still, I know the urgencies of forest life quite well. He settled in. He learned how to live in a different, I think, maybe, in a more perfect way than any human being – certainly in many thousands of years. Would I like to know what he knows? Absolutely. But that is not for us; that is not our destiny. What is that Biblical saying… "something that surpasseth understanding?" I would think you should be grateful that he has managed it. Happy for him. Like you mentioned, when we first spoke of it, there is no other bona fide example of this having ever happened to a human being. Not that we know of."

Lev was deeply quiet, inside. For most of his life he had thought about nature from a mostly comfortable, air-conditioned house with its ultimately silly backyard and picket fence.

Then he uttered, "We don't know the first thing about nature, do we, Taman?"

"What do you mean?"

"I mean other than all the carnage and entrails our species has meted out. And the odd bird feeder to dull the pain, or delude us into thinking that our capacity to be kind now and then makes up for everything else and presupposes the oldest allegories."

"Is there a moral to that story, David?"

"Absolutely. G-d preserved the Creation, for at least one good person."

The two men stood silently together, the loves of their lives beside them. Paris all around them.

"I suppose I envy him," David Lev said, almost rhapsodically. "He is so much better than I."

And Sasha witnessed her husband giving free vent to tears for the first time since she could remember, if ever.

© M.C. Tobias

GPSR Compliance

The European Union's (EU) General Product Safety Regulation (GPSR) is a set of rules that requires consumer products to be safe and our obligations to ensure this.

If you have any concerns about our products, you can contact us on

ProductSafety@springernature.com

In case Publisher is established outside the EU, the EU authorized representative is:

Springer Nature Customer Service Center GmbH
Europaplatz 3
69115 Heidelberg, Germany

www.ingramcontent.com/pod-product-compliance
Lightning Source LLC
LaVergne TN
LVHW010959250326
834688LV00003B/29